Fire and Rescue English Keywords

消防救援英语关键词

主编：王莉

副主编：史才华

中央编译出版社
CCTP Central Compilation & Translation Press

图书在版编目 (CIP) 数据

消防救援英语关键词 / 王莉, 史才华编 . —北京：
中央编译出版社，2023.10

ISBN 978-7-5117-4553-8

Ⅰ.①消… Ⅱ.①王… ②史… Ⅲ.①消防 – 救援 –
英语 Ⅳ.① TU998.1

中国国家版本馆 CIP 数据核字 (2023) 第 196283 号

消防救援英语关键词

出版统筹	张远航	
责任编辑	何 蕾	
责任印制	李 颖	
出版发行	中央编译出版社	
网 址	www.cctpcm.com	
地 址	北京市海淀区北四环西路 69 号 (100080)	
电 话	(010)55627391(总编室)	(010)55627116(编辑室)
	(010)55627320(发行部)	(010)55627377(新技术部)
经 销	全国新华书店	
印 刷	北京印刷集团有限责任公司印刷一厂	
开 本	880 毫米 × 1230 毫米 1/32	
字 数	98 千字	
印 张	6.875	
版 次	2023 年 10 月第 1 版	
印 次	2023 年 10 月第 1 次印刷	
定 价	39.80 元	

新浪微博：@中央编译出版社 微 信：中央编译出版社(ID: cctphome)
淘宝店铺：中央编译出版社直销店 (http: //shop108367160.taobao.com) (010)55627331

本社常年法律顾问：北京市吴栾赵阎律师事务所律师 闫军 梁勤
凡有印装质量问题，本社负责调换，电话：(010)55627320

序　言

　　本书旨在帮助国内外消防救援人员快速查阅国际消防救援常用的消防救援英语术语及与之对应的中国消防救援的中文术语，避免专业交流中产生歧义和误解。内容不仅包括中国国家标准化管理委员会发布的 GB/T 5907 部分常用消防词汇，还包括英国与美国的常用消防救援英语术语、消防救援单位与职务、消防救援衔、有限空间救援器材等英文译法。本书可供消防救援英语学习、国际消防救援交流以及所有英语爱好者学习使用。

　　非常感谢中国消防救援学院、国家消防救援局应急管理部同行在本书的审阅和修改过程中给予的宝贵帮助，特别是北京市消防救援总队昌平支队永安消防站的李士鹏站长和崔学伟副站长的专业支持，感谢英国消防学院等国际消防救援机构与所有同仁的大力支持。

编者

2023 年 10 月

Introduction

This book is to assist those studying English of fire & rescue to quickly consult commonly used fire rescue English terms and expressions in global fire & rescue service. It will also be of value to individuals in the fire sector, and to those undertaking induction training and some specialist roles in international fire and rescue communication. However, it is hoped that for whatever purpose it is used, it should act as a stimulus to further study and understanding of this ever developing discipline.

The contents are based largely upon standards and practices adopted in the PRC, the UK and the USA.

With this brief introduction, I would like to acknowledge the invaluable assistance given by the chiefs and my colleagues in China Fire and Rescue Institute in reviewing and revising the material for the publication, staff of the National Fire and Rescue Administration (NFRA), Ministry of the Emergency Management of the PRC, as well as members of the International Fire & Rescue Services, especially Chief Li Shipeng and Deputy Chief Cui Xuewei from Yong'an Fire Station, Changping Detachment, Beijing Fire & Rescue Corps of the NFRA.

Editor
October 2023

目　录
Contents

A

- **Absorption /əbˈzɔːpʃ(ə)n/:** the process of liquid, gas or other substance being taken in

  （液体、气体等的）吸收：吸收液体、气体或其他物质的过程。

- **Accelerant /əkˈselərənt/:** the flammable fuel (often liquid) used by some arsonists to increase size or intensity of fire; may also be accidentally introduced when a Hazmat becomes involved in fire

  助燃剂：纵火犯用以助长火势的可燃物（通常为液体）或火灾中意外出现的危险物品。

- **Accountability /əˌkaʊntəˈbɪləti/:** the process of emergency responders (from departments of fire, police, SAR, emergency medicine, etc.) checking into and making themselves announced as being on-scene during an incident to an incident commander or accountability officer. Through the accountability system, each person is tracked throughout the incident until released from the scene by the incident commander or accountability officer. This is becoming a standard in the emergency services arena primarily for the safety of emergency personnel, and it may implement a name tag system or personal locator device (a tracking device used by each individual that is linked to a computer).

  责任制/问责制: 在事故处理中,紧急救援人员(消防、警察、搜救、紧急医疗等)向指挥员或问责官员报到并陈述现场实情的过程。按照问责制度,指挥员和问责官员允许后,事故现场所有人员才可离开现场。问责制正在成为应急服务领域保障应急人员安全的标准。

问责系统可以采用姓名标签系统或个人定位装置（一种追踪装置，可与计算机联通）。

- **Acute toxicity/əˈkjuːt-tɔkˈsisəti/:** toxicity caused by exposure to a large dose (high concentration) for a short time (15min) or multiple exposure to a certain toxicant within 24 hours

  急性毒性： 短时间（15分钟）内一次性暴露于某种大剂量（高浓度）或24小时内多次暴露于某种小剂量（低浓度）毒物中所产生的毒性。

- **Acceptable fire risk /əkˈseptəb(ə)l - ˈfaiə - risk/:** in the fire risk assessment stage, the risk meets the prescribed acceptance criteria

  可接受火灾风险： 在火灾风险评估阶段，满足规定验收标准的风险。

- **Alarm /əˈlɑːm/:** (1) a system for detecting and reporting unusual conditions, such as smoke, fire, flood, loss of air, HAZMAT release, etc.; (2) a specific assignment of multiple fire companies and/or units to a particular incident, usually of fire in nature; (3) centralized dispatch center for interpreting alarms and dispatching resources

警报:（1）检测和报告异常情况的系统，如烟雾、火灾、洪水、缺氧、危险品释放等情况；（2）多个消防队和（或）特勤支队执行的特定任务，通常是自然火灾；（3）集中调度中心传达警报和调度资源。

- **Alarm of fire /əˈlɑːm - əv - ˈfaiə/:** an alarm issued by a person or an automatic device to notify the occurrence of a fire

  火灾警报: 人或自动装置通报火灾发生的警报。

- **Alarm time /əˈlɑːm-taɪm/:** the time when the fire alarm reception desk receives the report of fire alarm signal

  接警时间: 火警受理台接到火灾报警信号的时刻。

- **Ambulance /ˈæmbjʊl(ə)ns/:** a vehicle with special equipment to take sick or injured people to a hospital

救护车：配有特殊装备，用于将病人或伤者送往医院的车辆。

- **Apparatus /ˌæpəˈreɪtəs/:** a term usually used by firefighters to describe a vehicle (i.e. fire engine)

  灭火装备：消防员通常用来描述车辆（即消防车）的术语。

- **Area burning rate /ˈeərɪə-bɜːnɪŋ-reɪt/:** under the specified test conditions, the area of material burning per unit time

  面积燃烧速率：在规定的试验条件下，单位时间内材料燃烧的面积。

- **Attendance time /əˈtend(ə)ns-taɪm/:** the time from when a fire or other emergency signal is received to the time when the fire brigade arrives at the scene

  到场时间：接到火灾或其他紧急信号到消防队到达现场的时间。

- **Automatic control system for fire protection /ɔːtəˈmætɪk-kənˈtrəul-sɪstəm-fɔː-ˈfaɪə-prəˈtekʃ(ə)n/:** it usually consists of fire fighting linkage controllers, modules, gas fire-extinguishing controllers, fire fighting electrical control devices, emergency power supplies for fire fighting equipment, fire fighting emergency broadcast equipment, fire fighting telephones, transmission equipment, fire control center graphic display devices, fire fighting electric devices, fire hydrant buttons, etc. In the automatic fire alarm system, the fire alarm signal sent by the fire alarm controller is received to complete the control system of various fire fighting functions.

消防联动控制系统：通常由消防联动控制器、模块、气体灭火控制器、消防电气控制装置、消防设备应急电源、消防应急广播设备、消防电话、传输设备、消防控制中心图形显示装置、消防电动装置、消火栓按钮等设备组成，在火灾自动报警系统中，接收火灾报警控制器发出的火灾报警信号，完成各项消防功能的控制系统。

- **Automatic sprinkler system /ɔːtəˈmætɪk-ˈsprɪŋklə-ˈsɪstəm/:** it is composed of sprinkler, alarm valve group, water flow alarm device (water flow indicator or pressure switch) and other components, as well as pipelines, water supply facilities, and an automatic fire extinguishing system that can spray water in the event of a fire

  自动喷水灭火系统: 由洒水喷头、报警阀组、水流报警装置(水流指示器或压力开关)等组件,以及管道、供水设施组成,并能在发生火灾时喷水的自动灭火系统。

B

- **Backdraft /ˈbækˌdrɑːft/:** a fire phenomenon caused when heat and heavy smoke (unburned fuel particles) accumulate inside a compartment, depleting the available air, and then oxygen/air is re-introduced, completing the *fire triangle* and causing rapid combustion

  回燃：当热量和浓烟（未燃烧的燃料颗粒）在受限空间内积聚，耗尽了可用的空气，氧气突然引入该密闭空间而导致的爆燃。

- **Base injection foam extinguishing system /beɪs-inˈdʒekʃ(ə)n-fəum - ikˈstiŋgwiʃiŋ - ˈsɪstəm/:** it can inject foam under the surface of combustible liquid, and the foam rises to the surface of the liquid and then spreads to form a foam extinguishing system

液下喷射泡沫灭火系统: 在可燃液体表面下注入泡沫，泡沫上升到液体表面并扩散开，形成泡沫层的泡沫灭火系统。

- **Balcony /ˈbælkənɪ/:** a platform with a wall or railing built onto the outside wall of a building

  阳台: 建筑物外墙上建有墙或栏杆的平台。

- **Beam /biːm/:** a long thick bar of wood, metal, or concrete, especially one used to support the roof of a building

  梁: 由木头、金属或混凝土制成的长而粗的杆，尤指用来支撑建筑物屋顶的横梁。

- **Blanket /ˈblæŋkɪt/:** cover something completely

  覆盖: 完全盖住某物。

- **Blow through mechanical pressurization /blǝʊ - θruː - mɪˈkænɪk(ǝ)l - ˌprɛʃǝrɪˈzefǝn/:** the way that the fan directly pressurizes the stairwell mechanically without passing through mechanical pressurization

  直灌式加压送风： 风机未通过送风井道直接对楼梯井机械加压送风的方式。

- **Boil over /bɒɪl - ˈǝʊvǝ/:** a phenomenon in which the water layer under the burning oil layer boils and expands due to heating, causing the burning oil to splash, which causes the combustion to increase instantly

  沸溢： 正在燃烧的油层下的水层因受热沸腾膨胀导致燃烧着的油品喷溅，使燃烧瞬间增大的现象。

- **Fire branch /ˈfaiǝ - brɑːn(t)ʃ/**

  消防枪

- **Branch man /brɑːn(t)ʃ - mæn/:** a fireman who controls a nozzle

  水枪手： 控制水枪的消防员。

- **Circuit-breaker /ˈsɜːkɪt-breɪkə(r)/:** a device which can stop the flow of electricity around a circuit by switching itself off if anything goes worng

  断路器： 在电路出现故障时自行关闭，从而停止电路周围电流的装置。

- **Breathing Apparatus (BA) /ˈbriːðɪŋ-ˌæpəˈreɪtəs/:** respiratory equipment

  呼吸器： 呼吸设备。

- **Bridgehead /ˈbrɪdʒhed/**

  进攻起点；水枪阵地

- **Buckle /ˈbʌkl/:** a piece of metal or plastic attached to one end of a belt or strap, which is used to fasten it

  （皮带等的）带扣： 系在皮带或带子一端的金属或塑料，用来系紧皮带或带子。

C

- **Carbon dioxide extinguishing system** /ˈkɑːb(ə)n - daɪˈɒksaɪd - ikˈstiŋgwiʃiŋ-ˈsɪstəm/: gas fire extinguishing system consisting of carbon dioxide supply source, nozzles and pipelines

  二氧化碳灭火系统：由二氧化碳供应源、喷嘴和管路等组成的气体灭火系统。

- **Cable** /ˈkeibl/: a set of wires, covered by plastic or rubber, which carries electricity, telephone signals, etc.

  电缆：一组覆盖着塑料或橡胶的电线，用来传输电力、电话信号等。

- **Cardiopulmonary resuscitation /ˌkɑːdɪəʊˈpʌlmən(ə)rɪ- rɪˌsʌsɪˈteɪʃən/:** to restore people's breathing and heartbeat through manual intervention

  心肺复苏：通过人工干预手段，使人的呼吸和心跳恢复正常。

- **Centralized receipt of fire alarm /ˈsentrəlaizd - rɪˈsiːt-əv - ˈfaɪə - əˈlɑːm/:** after the fire general and branch (big) brigade collectively accept the report of fire alarm, they then issue a combat command to the squadron in the jurisdiction

  集中接警：消防总、支(大)队集中受理火灾报警后，再向辖区中队发出战斗指令的接警方式。

- **Charge a hose /tʃɑːdʒ - ə - həʊz/:** to make water pressure available on a hose in final preparation for its use. This is done on the scene after the hose is deployed.

  水带加压：水带部署好后，在现场为水带提供水压，为其使用做最后准备。

- **Char /tʃɑ:/:** a carbon-containing residue formed during pyrolysis or incomplete combustion of a substance

  炭：物质在热解或不完全燃烧过程中形成的含碳残余物。

- **Char length /tʃɑ: - leŋθ/:** under the specified test conditions, the maximum length of carbonization of the material in a specific direction

  炭化长度：在规定的试验条件下，材料在特定方向上发生炭化的最大长度。

- **Chimney effect /ˈtʃimni-iˈfekt/:** in a relatively closed vertical space, the phenomenon of upward flow of smoke and hot air due to air convection

  烟囱效应：在相对封闭的竖向空间内，由于气流对流而使烟和热气流向上流动的现象。

- **Chronic toxicity /ˈkrɒnɪk - tɔkˈsisəti/:** toxicity caused by exposure to small doses (low concentrations) of a certain toxicant for many times for a long time

慢性毒性：长时间内多次暴露于某种小剂量（低浓度）毒物中所产生的毒性。

- **Circulation /sɜːkjʊˈleɪʃ(ə)n/:** the movement of something (for example air, water, gas, etc.) around an area or inside a system or machine

  （气、水等的）环流，循环：某物（如空气、水、煤气等）在一个区域周围、系统或机器内部的运动。

- **Class A fire /klɑːs-ei-faiə/:** a fire involving combustibles such as wood, paper, and other natural materials

  A 类火灾：可燃物引起的火灾，如木材、纸张和其他天然材料。

- **Class B fire /klɑːs-bi-faiə/:** a fire involving hydrocarbons

  B 类火灾：碳氢化合物引起的火灾。

- **Class C fire /klɑːs-si-faiə/:** an electrical fire

  C 类火灾：电气引起的火灾。

- **Class D fire /klɑːs-di-faiə/:** a fire involving metals, such as sodium, titanium, magnesium, potasium, uranium, lithium, plutonium and calcium

  D 类火灾：金属引起的火灾，如钠、钛、镁、钾、铀、锂、钚和钙。

- **Class F fire /klɑːs-ev-faiə/:** it refers to the fire of cooking materials (such as animal and vegetable oils) in cooking utensils

  F 类火灾：烹饪器具内的烹饪物引起的（如动植物油脂）火灾。

- **Cockloft /ˈkɒklɒft/:** a structural space above ceilings and below rafters, often connecting adjacent occupancies and permitting fire to spread laterally, often unseen

  阁楼：天花板上方和椽子下方的结构空间，通常连接相邻的建筑，可使火势横向蔓延，通常不易察觉。

- **Collapse zone /kəˈlæps-zəʊn/:** the area around a structure that would contain debris if the building collapsed

塌陷区：建筑物倒塌后的残骸区。

- **Collision /kəˈlɪʒ(ə)n/:** strong disagreement; conflict

  相撞，碰撞：强烈的抵触；冲突

- **Company /ˈkʌmp(ə)nɪ/:** a team organized by two or more firefighters, led by a fire officer, and equipped to perform certain operational functions

  （消防）队：两名（含）以上消防员组成一个团队，由一名指挥员领导，配备执行特定行动任务的装备。

- **Combustion /kəmˈbʌstʃ(ə)n/:** the process of burning something; a chemical process in which substances combine with the oxygen in the air to produce heat and light

  燃烧：燃烧过程；可燃物与空气中的氧放出热和光的放热过程。

- **Commander /kəˈmɑːndə/:** a person who commands

指挥员：下达命令的人员。

- **Communication** /kəmjuːniˈkeiʃ(ə)n/
  通信

- **Composition** /kɒmpəˈziʃ(ə)n/: the different parts of something
  成分

- **Company officer** /ˈkʌmp(ə)nɪ - ˈɒfɪsə/: a fire officer
  队长：消防指挥员。

- **Compartment fire** /kəmˈpɑːtm(ə)nt - ˈfaɪə/: an "isolated" fire, or a fire which is "boxed in" or "closed off" from the rest of the structure. An example of this is a fire in a room where all the windows and doors are closed preventing the fire from spreading to other rooms.
  隔间火灾："孤立的"火灾，或与建筑物的其他部分"隔开"或"封闭"的火灾。例如，为防止火势蔓延到其他房间，关闭起火房间的所有门窗，令燃烧范围仅此房间的火灾。

- **Configuration /kənˌfɪɡəˈreɪʃ(ə)n/:** an arrangement of a group of things

  布局：一组事物的排列。

- **Confined space /kənˈfaɪnd - speɪs/:** a confined space is any space: (1) has limited or restricted means of entry or exit; (2) is large enough for a person to enter to perform tasks; (3) and is not designed or configured for continuous occupancy

  密闭空间：（1）进出方式有限或受限的空间；（2）仅可容纳一人进入执行任务的空间；（3）不是为连续使用而设计或配置的空间。

- **Constituent /kənˈstɪtʃuənt/:** one of the parts of something that combine to form the whole

  组成部分：组成整体的部分。

- **Convection /kənˈvekʃ(ə)n/:** the process in which heat

moves through a gas or a liquid as the hotter part rises
and the cooler, heavier part sinks

（热通过气体或液体的）运流，对流：热量在气体或
液体中流动的过程，热的部分上升，冷的、重的部分
下沉。

- **Cordon /ˈkɔː:d(ə)n/** : a line or ring of police, soldiers, or
  vehicles preventing people from entering or leaving an
  area

  火场警戒线：由警察、士兵或车辆等形成的区域，避
  免进出。

- **Cross lay /krɒs - leɪ/:** an arrangement of hoses on a
  pumper so that it can be quickly unloaded from either side
  of the apparatus; often pre-connected to a pump outlet
  and equipped with a suitable nozzle. It is also known as
  Mattydale Lay.

  交叉铺设：在泵车上布置软管，使其可以从设备的任
  何一侧快速卸下；通常预先连接到泵的出口，并配备
  合适的喷嘴。

- **Collapse /kəˈlæps/:** falls down very suddenly

  坍塌：突然塌落。

- **Collide /kəˈlaɪd/:** to come together with solid or direct impact

  碰撞，撞击

- **Column /ˈkɒləm/:** something that has a tall, narrow shape

  柱状物：形状高而细的物体，消防用语中可以用 a column of smoke，指烟柱。

- **Compound /ˈkɒmpaʊnd/:** a substance formed by a chemical reaction of two or more elements in fixed amounts relative to each other

  化合物：两种或两种以上元素以相对固定的量进行化学反应形成的物质。

- **Combustibility /kəmˌbʌstəˈbiliti/:** under the specified test conditions, the material can be ignited and can continue to combust

可燃性：在规定的试验条件下，物质能够被引燃且能持续燃烧的特性。

- **Combined distribution system /kəmˈbaɪnd - dɪstrɪˈbjuːʃ(ə)n - ˈsɪstəm/:** use a set of fire extinguishing agent storage device to protect two or more protective areas or gas fire extinguishing systems of the protected objects

组合分配系统：使用一套灭火剂贮存装置，保护两个及以上防护区或保护对象的气体灭火系统。

- **Command post on the fire ground /kəˈmɑːnd - pəʊst - ɒn - ðə - ˈfaɪə - graʊnd/:** when fighting a fire, in order to coordinate the fire fighting and rescue operations of the fire fighting and rescue forces on the fire ground, the implementation of unified organization, unified command and unified action, a temporary command organization composed of relevant personnel

火场指挥部：扑救火灾时，为协调灭火救援力量在火场上的灭火救援行动，实施统一组织、统一指挥、统一行动，由相关人员组成的临时指挥机构。

- **Conduction /kənˈdʌkʃ(ə)n/:** the process by which heat or electricity passes through a material

  （热或电等能量的）传导：热量或电通过材料的过程。

- **Construction /kənˈstrʌkʃ(ə)n/:** the building of things such as houses, factories, roads, and bridges, etc.

  建造：建造房屋、工厂、道路和桥梁等。

- **Continuous /kənˈtɪnjʊəs/:** happening or existing for a period of time without interruption

  持续的：过程或事件不间断的延续一段时间。

- **Current /ˈkʌr(ə)nt/:** the flow of electricity through a wire, etc.

  电流：电能（电子）通过电线等导体的流动。

- **Cylinder /ˈsɪlɪndə/**: (1) a solid or hollow figure with round ends and long straight sides; (2) the hollow tube in an engine, shaped like a cylinder

  （1）圆柱（体），圆柱状物；（2）（发动机的）汽缸

D

- **Damp down /dæmp - daʊn/:** after the fire is extinguished, measures such as watering shall be taken to eliminate the risk of reignition of possible hidden fumes and heating residues

  消除残火：火灾扑灭后，给火灾现场洒水，以避免隐藏的烟雾和加热残留物引起复燃。

- **Debriefing /diˈbriːfɪŋ/:** a meeting where someone such as a soldier, diplomat, or astronant is asked to give a report on an operation or task that they have just completed

任务报告，汇报：士兵、外交官或宇航员就刚刚完成的行动或任务所做的报告。

- **Deflagration /ˌdeflə'greiʃən/:** the combustion wave propagating at subsonic speed

  Note: If in a gas medium, deflagration is the same as flame.

  爆燃：以亚音速传播的燃烧波。

  注：在气体介质内，爆燃就是火焰。

- **Dead end /ded - end/:** a road, passage, etc., that is closed at one end

  袋形走道：一头封死的道路（或通道等）。

- **Debris /'debri:; də'bri/:** scattered fragments; wreckage

  残骸：散落的碎片；残片。

- **Deformation /ˌdi:fɔː'meɪʃ(ə)n/:** the act of deforming; distortion

变形：改变形状的行为；扭曲。

- **Deluge extinguishing system /ˈdeljuːdʒ - iksˈtiŋgwiʃiŋ - ˈsistəm/:** an open automatic sprinkler system that is controlled by the automatic fire alarm system or transmission pipe, automatically opens the rain alarm valve and starts the water supply pump, and then supplies water to the open sprinkler

  雨淋灭火系统：由火灾自动报警系统或传动管控制，自动开启雨淋报警阀和启动供水泵后，向开式洒水喷头供水的开式自动喷水灭火系统。

- **Defense attack /dɪˈfensɪv - əˈtæk/:** a primarily exterior attack form often used when fighting the fire directly or from within a structure is not feasible due to dangers from direct flame, heat, structural collapse or the presence of hazardous materials. Often structures which are fully involved are attacked defensively with the main goal of the protection of nearby exposures. This form of attack is far less effective than an offensive or direct attack.

防御型灭火作战：主要是从外部灭火的战术，通常在直接灭火或内部灭火不可行的情况下使用，因为存在火焰、热量、建筑物坍塌或危险材料的威胁。通常情况下，对完全燃烧的建筑物进行防御性灭火，主要目标是保护附近的暴露物。这种形式远不如进攻型灭火或直接抑制有效。

- **Dedicated line for dispatching fire fighting force /ˈdedɪˌkeɪtɪd - laɪn- fɔ - dɪˈspætʃ - ˈfaɪə -ˈfaɪtɪŋ - fɔːs/:** wire fire communication line used to dispatch fire fighting forces

火警调度专线：用于调度灭火力量的消防有线通信线路。

- **Detonation /ˌdetnˈeɪʃən/:** a chemical reaction characterized by shock waves that propagate faster than the speed of sound in unreacted substances

爆轰：以冲击波为特征，在未反应物质中传播速度大于声速的化学反应。

- **Design fire /dɪˈzaɪn - ˈfaɪə/:** a quantitative description of assumed fire characteristics for a design fire scenario

  设定火灾：对一个设定火灾场景假定火灾特征的定量描述。

- **Design fire scenario /dɪˈzaɪn - ˈfaɪə - sɪˈnɑːrɪəʊ/:** specific fire scenarios used for deterministic fire safety engineering analysis

  Note: Because there are many possible fire scenarios, it is necessary to select the most important scene (a set fire scene) for analysis. The choice of setting fire scenes is compatible with the fire safety design goals, and can explain the possibility and consequences of potential fire scenarios.

  设定火灾场景：进行确定性消防安全工程分析所采用的特定火灾场景。

  注：因为可能的火灾场景非常多，所以有必要选择最重要的场景（设定火灾场景）进行分析。设定火灾场景的选择是和火灾安全设计目标相适应的，并且能说明潜在火灾场景的可能性和后果。

- **Delayed toxicity /diˈleid - tɔkˈsisəti/:** toxicity occurs following a period of incubation period after stopping exposure to a certain toxicant

  潜伏毒性：停止接触某种毒物后，过一段时间才出现的毒性。

- **Delivery hose /dɪˈlɪv(ə)rɪ - həʊz/**

  供水水带

- **Density /ˈdensɪtɪ/:** the thickness of a solid, liquid or gas measured by its mass per unit volume

  密度：固体、液体或气体的密实度，以其单位体积的质量来衡量。

- **Direct attack /dɪˈrekt - əˈtæk/:** Putting the wet stuff on the burning stuff. A form of fire attack in which hoses are advanced to the fire inside a structure and hose streams are directed at the burning materials.

  直接灭火："把灭火剂直接放在燃烧物上。"一种灭火方式，将水带运进火场内部，水流直接喷向燃烧物质。

- **Dispatch /dis'pætʃ/:** refers to some specific person or place designated for handling a call for help by alerting the specific resources necessary

  调度：指的是指定的人或地点，通过提醒必要的特定资源来处理求助电话。

- **Distributor /dɪ'strɪbjʊtə/**

  配电器；分水器

- **Direct transmission /dɪ'rekt - trænz'mɪʃ(ə)n/:** between the fire ground TV transmitter and the receiving location, the live situation of the fire site is directly transmitted to the receiving site without any switching equipment

  直接传输：火场电视发射机与接收地点之间不经过任何转接设备将火场实况直接传送到接收地点的传输方式。

- **Deterioration of hose /dɪˌtɪərɪə'reɪʃən - əv - həʊz/**

  水带损坏

- **Diffusion /dɪˈfjuːʒ(ə)n/:** the process whereby particles of liquids, gases, or solids intermingle as the result of their spontaneous movement caused by the thermal agitation and dissolved substances move from a region of higher to one of lower concentration

  扩散：液体、气体或固体颗粒由于热搅动引起的自发运动而混合在一起的过程，溶解物质从浓度较高的区域移动到浓度较低的区域。

- **Difficult flammability /ˈdɪfɪk(ə)lt - ˌflæməˈbɪləti/:** under the specified test conditions, the material is difficult to be ignited

  难燃性：在规定的试验条件下，物质难以进行有焰燃烧的特性。

- **Dutch roll /dʌtʃ - rəʊl/**
  （消防）对折卷（水带）

- **Dry pipe automatic sprinkler system /draɪ - paɪp - ɔːtəˈmætɪk - ˈsprɪŋklə - ˈsɪstəm/:** in the quasi-working state, the water distribution pipeline is filled with a sealed automatic sprinkler system with pressurized gas used to start the system

干式自动喷水灭火系统：准工作状态时，配水管道内充满用于启动系统的有压气体的闭式自动喷水灭火系统。

• **Drencher extinguishing system /ˈdrɛntʃɚ - iksˈtiŋgwiʃiŋ - ˈsɪstəm/:** it is composed of an open sprinkler or water curtain sprinkler, a rain alarm valve or a temperature-sensitive rain alarm valve, a water flow alarm device (water flow indicator or pressure switch), etc., which is an open automatic sprinkler system used to block smoke and fire, and cool the partition

水幕灭火系统：由开式洒水喷头或水幕喷头、雨淋报警阀或感温雨淋报警阀、水流报警装置（水流指示器或压力开关）等组成，用于挡烟阻火和冷却分隔物的开式自动喷水灭火系统。

• **Dispatch of fire fighting force /disˈpætʃ - ɒv - ˈfaɪə -**

ˈfaɪtɪŋ - ˈfɔ:s/: the process of dispatching fire fighting and rescue forces to the fire ground through the fire communication command system after receiving the fire alarm

力量调度：接警后，消防通信指挥系统向火场调派消防救援力量的过程。

- **Duration of fire resistance** /djʊˈreɪʃ(ə)n - əv - ˈfaɪə - rɪˈzɪst(ə)ns/: under standard fire test conditions, building components, fittings or structures start from the time of being exposed to fire to the time of losing their fire stability, fire integrity or fire insulation

耐火极限：在标准耐火试验条件下，建筑构件、配件或结构从受到火的作用时起，到失去支持能力、完整性被破坏或失去隔火作用为止的时间。

- **Dust explosions** /dʌst - ɪkˈspləʊʒ(ə)n/: the fast combustion of dust particles suspended in the air in an enclosed location. Coal dust explosions are a frequent hazard in underground coal mines, but dust explosions can occur where any powdered combustible material is present in an enclosed atmosphere or, in general, in high enough concentrations of dispersed combustible particles in atmosphere or other suitable gaseous medium such as molecular oxygen

粉尘爆炸：粉尘爆炸是悬浮在空气中的粉尘颗粒在密闭场所内快速燃烧。煤气爆炸是煤矿井下常见的危险，但粉尘爆炸可能发生在任何粉末状可燃物质，或者通常在大气或其他适宜的气体介质（如氧分子）中分散的可燃颗粒浓度足度高的情况下。

E

- **Effort /ˈefət/:** effective force as distinguished from the possible resistance called into action by such a force

  力点：有效力量，区别于此力量可能引起的阻力。

- **Elasticity /elæˈstɪsɪtɪ/:** the ability of material or substance to return to its original shape, size, and condition after it has been stretched

  弹性：材料或物质在拉伸后能够恢复到其原始形状、尺寸和状态的能力。

- **Electrical reciprocating saws /ɪˈlektrɪk(ə)l - riˈsɪprəkeitiŋ - sɔːz/**
  电动往复锯

- **Element /ˈelɪm(ə)nt/:** a substance such as gold, oxygen, or carbon that consists of only one type of atom
  元素：仅由一种原子组成的物质，如金、氧或碳。

- **EMS:** Emergency Medical Service(s)
  紧急医疗服务

- **Emergency staircase /iˈmɜːdʒənsi - ˈsteəkeɪs/:** an indoor or outdoor staircase with sufficient fire resistance and used as a vertical evacuation access
  疏散楼梯：具有足够防火能力并作为竖向疏散通道的室内或室外楼梯。

- **Energy /ˈenədʒɪ/:** the ability of matter or radiation to work
  能；能量

- **Fire ergine /ˈfaɪə-ˈendʒzn/:** a fire suppression vehicle that has a water pump and, typically, is designed to carry fire hoses and a limited supply of water

  泵浦消防车: 有水泵的灭火车辆,通常携带消防水带,提供有限的水量。

- **Engineer /endʒɪˈnɪə/:** a firefighter responsible for driving the engine to the scene of the call and operation of the pumps on an engine, to provide sufficient water to the hose. The term could be either a position title or a rank; usage varies among departments.

  工程师: 负责驾驶消防车至报警现场,操作消防车载水泵,为水带供水。该词既可表示职务名称,也可表示级别;各部门的用法各不相同。

- **Engine company /ˈendʒɪn - ˈkʌmp(ə)nɪ/:** a group of firefighters assigned to an apparatus with a water pump and equipped with fire hoses and other tools related to fire extinguishment

  泵浦车中队: 泵浦消防车配备的消防员,该车配有水泵、消防水带和其他灭火工具。

- **Engine pressure /ˈendʒɪn-ˈpreʃə/:** the pressure in a fire hose measured at the outlet of the pump

  水压： 在水泵出口处测量的消防水带的压力。

- **Escalate /ˈeskəleɪt /:** develope; increase; become more intense

  升级： 发展；增加；变得更加严重。

- **Escape route /ɪˈskeɪp - ruːt/:** the way to reach the safety exit in case of an emergency

  疏散路线： 紧急情况下，到达安全出口的路径。

- **Escape access /ɪˈskeɪp - ˈækses/:** a passage may in a building with sufficient fire and smoke preventing capacity which meets the requirements safe evacuation of personnel

  疏散通道： 建筑物内具有足够防火和防烟能力，主要满足人员安全疏散要求的通道。

- **Evacuation plan /ɪˌvækjʊˈeɪʃ(ə)n - plæn/:** a plan made in advance to ensure that people in the building can evacuate safely in the event of a fire

  疏散预案： 为保证建筑物内人员在火灾情况下能安全疏散而事先制定的计划。

- **Evaporation /iˌvæpəˈreiʃən/:** the process of a liquid changing into a gas

  蒸发： 液体变成气体的过程。

- **Emergency lighting /iˈməːdʒənsi - ˈlaɪtɪŋ/:** When normal lighting is interrupted, it is used for evacuation and lighting for fire fighting operations.

  应急照明： 正常照明中断时，用于人员疏散和消防作业的照明。

- **Electrical fire monitoring system /ɪˈlektrɪk(ə)l - ˈfaɪə - ˈmɒnɪtərɪŋ - ˈsɪstəm/:** it is composed of electrical fire monitoring equipment and electrical fire monitoring detectors. When the detected parameters in the protected electrical circuit exceed the alarm set value, the system can send out alarm signals, control signals and indicate the alarm position.

电气火灾监控系统：由电气火灾监控设备、电气火灾监控探测器组成，当被保护电气线路中的被探测参数超过报警设定值时，能发出报警信号、控制信号并能指示报警部位的系统。

- **Enclosed staircase /ɪnˈklozd - ˈsteəkeɪs/:** use two-way spring doors, fire doors and other measures to separate the stairwell that can prevent fire smoke and heat from entering

封闭楼梯间：采用双向弹簧门、防火门等分隔的楼梯间，能防止火灾的烟气和热气进入。

- **Emergency lighting system /iˈmɜːdʒənsi - ˈlaɪtɪŋ - ˈsɪstəm/:** lamps and related devices used for emergency lighting

应急照明系统：用于应急照明的灯具及相关装置。

- **Escape direction sign /ɪˈskeɪp - dəˈrekʃn - saɪn/:** signs set on the safety exit and evacuation route are used to indicate the safety exit and the route leading to the safety exit

  Note: The evacuation sign is the "safe exit" sign or the combination of the "safe exit" and "evacuation passage direction" signs in GB 13495. GB 13495 specifies the style of the logo and the style of combination.

  疏散指示标志：设置在安全出口和疏散路线上，用于指示安全出口和通向安全出口的路线。

  注：疏散指示标志是 GB 13495 中的"安全出口"标志或"安全出口"与"疏散通道方向"标志的组合。GB 13495 规定了标志的式样以及组合使用的式样等内容。

- **Evacuation time /ɪˌvækjʊˈeɪʃ(ə)n - taɪm/:** the time from when all people in the building or a certain area of the building receive the fire information to when they arrive at the safety exit or safe zone

  疏散时间：建筑物内的所有人员从获得火灾信息至抵达安全出口或安全区的时间。

- **Evacuation /ɪˌvækjʊˈeɪʃ(ə)n/:** removal of personnel from a dangerous area, in particular, a HAZMAT incident, burning building, or other emergency; it also refers to act of removing firefighters from a structure in danger of collapsing

  疏散：将人员撤离危险区域，尤其是危险品事件、燃烧的建筑物或其他紧急情况，也指将消防员从有倒塌危险的建筑中撤出的行为。

- **Evacuation distance/ɪˌvækjʊˈeɪʃ(ə)n - ˈdɪst(ə)ns/:** the distance from any point in a room to the nearest emergency exit

  疏散距离：从房间内任一位置到最近的安全出口的距离。

- **Expand /ɪkˈspænd/:** to become greater in size, number or importance; to make something greater in size, number or importance

  扩大，增加，增强（尺码、数量或重要性）

- **Explosion limit /ɪkˈspləʊʒ(ə)n - ˈlɪmɪt/:** combustible gas, steam or dust is uniformly mixed with air to form a mixture, which will produce the highest or lowest concentration of explosion when there is enough ignition energy

  爆炸极限：可燃气体、蒸气或粉尘与空气均匀混合后形成混合气，遇到足够的点火能量会产生爆炸的最高或最低浓度。

- **Exposure hazard /ɪkˈspəʊʒə - ˈhæzəd/:** dangers caused by exposure to toxic gases or fire effluents

  暴露危险：暴露于有毒气体或火灾流出物环境中产生的危险。

- **Exposure /ɪkˈspəʊʒə/:** a property near fire that may become involved by transfer of heat or burning material from main fire, typically by convection or radiation, and may range from 40 feet (12 m) to several miles, depending on the size and type of fire or explosion

  暴露：火灾附近的财产可能因主火源的热量或燃烧物

质的传递而受到威胁，通常是通过对流或辐射。范围可能是 40 英尺（12 米）到几英里，取决于火灾或爆炸的级别和类型。

- **Exterior attack /ɪkˈstɪərɪə - əˈtæk/:** a method of extinguishing a fire which does not involve entering the structure, also known as surround and drown
  外攻：不进入结构内部的灭火方法，也称为包围式灭火。

- **Extrication /ˌekstriˈkeɪʃən/:** removal of a trapped victim such as vehicle extrication, confined space rescue, or trench rescue; sometimes using hydraulic spreaders, jaws of life, or other technical equipment
  解救：在车辆解救、密闭空间救援或沟渠救援时，移出被困人员；有时使用液压扩张器、救生钳或其他技术设备。

- **Extrication equipment /ˌekstriˈkeɪʃən - ɪˈkwɪpm(ə)nt/**
  解救设备

- **Explosion suppression** /ɪkˈspləʊʒ(ə)n - səˈpreʃ(ə)n/: a technology that automatically detects the occurrence of an explosion, extinguishes flames and suppresses the development of an explosion through physical and chemical effects

  抑爆：自动探测爆炸的发生，通过物理化学作用扑灭火焰、抑制爆炸发展的技术。

- **Extinguishing agent** /iksˈtɪŋgwiʃɪŋ - ˈeɪdʒ(ə)nt/: a substance that can effectively destroy the conditions of combustion and terminate combustion

  灭火剂：能够有效地破坏燃烧条件，停止燃烧的物质。

- **Exit passageway** /ˈeksɪt - ˈpæsɪdʒweɪ/: to set up smoke-proof facilities and separated by fire walls on both sides for safe passage of personnel to outdoor walkways

  出口通道：具有防烟设施且两侧采用防火墙分隔的走道，用于人员安全通行至室外。

F

- **FDC (Fire Department Connection) /ˈfaɪə - dɪˈpaːtmənt-kəˈnekʃn/:** the location in which pumping apparatus hooks to the standpipe and sprinkler system of a building

  消防水泵接合器：该接合器的接口用于将消防车水泵与建筑物的竖管和喷水灭火系统相连接。

- **Fire flow /ˈfaɪə - fləʊ/:** the amount of water being pumped onto a fire, or required to extinguish a fire, a critical calculation in light of the axiom that an ordinary fire will not be extinguished unless there is sufficient water to remove the heat of the fire

  灭火用水量：估算灭火所需泵入火场的水量。根据"除非有足够的水把火的热量带走，否则普通的火不会熄灭"而进行的重要计算。

- **Fire ground /ˈfaɪə - graʊnd/:** the operational area at the scene of a fire; area in which incident commander is in control. Also used as name of radio frequency to be used by units operating in the fire ground, as in "Responding units switch to fire ground."

  火场： 火灾现场的行动区域；事故指挥员的控制区域。也用作在火场各单位使用的无线电频道名称，如"各响应单位切换到火场（频道）。"

- **Fire inspector/ˈfaɪə - ɪnˈspektə/:** a person who is responsible for issuing permits and enforcing the fire code, including any necessary inspection of the premises, as before allowing (or during) a large indoor gathering

  消防检查员： 负责发放许可证和执行消防法规的人员，包括对场所进行一切必要的检查，如大型室内聚会之前（或过程中）的场所检查。

- **Fire load /ˈfaɪə - ləʊd/:** an estimate of the amount of heat that will be given off during ordinary combustion of all the fuel in a given space, e.g., a bedroom or a lumberyard

火灾负荷：对某一特定空间内所有燃料正常燃烧时产生的热量的估计，例如，一间卧室或一家木材厂。

- **Fire marshal /ˈfaɪə - ˈmɑːʃ(ə)l/**: an administrative and investigative office for fire prevention and arson investigation which has legal authority to enforce state and local fire laws in the USA

  消防队长：负责防火和纵火调查办公室的行政和调查事务，在美国拥有执行州和地方消防法的法律权力。

- **Fire protection /ˈfaɪə - prəˈtekʃ(ə)n/**: a collective term for fire prevention and fire fighting and rescue

  消防：火灾预防和灭火救援的统称。

- **Fire /ˈfaɪə/**: the phenomenon of combustion characterized by the release of heat accompanied by smoke or flame or both

  火：释放热量并伴有烟或火焰或两者兼有的燃烧现象。

- **Fire arson /ˈfaɪə - ˈɑːs(ə)n/**: the act of deliberately creating a fire

纵火：故意制造火灾的行为。

- **Fire parameter /ˈfaɪə - pəˈræmɪtə/:** a physical quantity that represents the characteristics of a fire

 火灾参数：表示火灾特性的物理量。

- **Fire classification /ˈfaɪə - ˌklæsɪfɪˈkeɪʃ(ə)n/:** according to the type of combustibles and the characteristics of combustion, the classification of fire which is carried out according to a standardized method

 Note: GB/T 4968 specifies the specific fire classification.

 火灾分类：根据可燃物的类型和燃烧特性，按标准化的方法对火灾进行的分类。

 注：GB/T 4968 规定了具体的火灾分类。

- **Fire mechanism /ˈfaɪə - ˈmek(ə)nɪz(ə)m/:** the physical and chemical laws of fire phenomenon

 火灾机理：与火灾现象相关的物理和化学规律。

- **Fire science /ˈfaɪə - ˈsaɪəns/:** the discipline that studies the mechanism, laws, characteristics, phenomena, and processes of fire

  火灾科学： 研究火灾机理、规律、特点、现象和过程等的学科。

- **Fire test /ˈfaɪə - test/:** the scientific experiments conducted to understand and explore the mechanism, laws, characteristics, phenomena, effects and processes of fire

  火灾试验： 为了解和探求火灾的机理、规律、特点、现象、影响和过程开展的科学试验。

- **Fire fighting /ˈfaɪə - ˈfaɪtɪŋ/:** the activities and processes of extinguishing or suppressing a fire

  灭火： 扑灭或抑制火灾的活动和过程。

- **Fire fighting technology /ˈfaɪə - ˈfaɪtɪŋ - tekˈnɒlədʒɪ/:** the collective term for scientific methods, materials, equipment, facilities, etc., used to extinguish fires

  灭火技术： 扑灭火灾的科学方法、材料、装备和设施等的统称。

- **Fire fighting and rescue /ˈfaɪə - ˈfaɪtɪŋ - ənd - ˈreskjuː/:** the fire fighting and implement rescue activities at the scene of fire focusing on saving lives

  灭火救援: 扑灭火灾以及在火灾现场抢救人员生命的援救活动。

- **Fire-extinguishing time /ˈfaɪə - iksˈtiŋgwiʃiŋ - taɪm/:** under the specified conditions, the time from the fire extinguishing device releasing the extinguishing agent to the complete extinguishment of the flame

  灭火时间: 在规定条件下，灭火装置释放灭火剂至火焰完全熄灭的时间。

- **Fire safety sign /ˈfaɪə - ˈseɪftɪ - saɪn/:** it is composed of graphic symbols, safety colors, geometric shapes (or borders), etc., which represents specific fire safety information, supplemented by safety signs with text or directions when necessary

Note: GB 13495 specifies specific fire safety signs.

消防安全标志： 由表示特定消防安全信息的图形符号、安全色、几何形状（或边框）等构成，必要时辅以文字或方向指示的安全标志。

注： GB 13495 规定了具体的消防安全标志。

- **Fire facility /ˈfaɪə - fəˈsɪləti/:** the fixed fire fighting systems and equipment, specially used for fire prevention, fire alarm, fire fighting and evacuation in case of fire, such as automatic fire alarm system, automatic fire extinguishing system, fire hydrant system, smoke exhaust system, emergency broadcast and emergency lighting, separation facilities and safety evacuation facilities

消防设施： 专门用于火灾预防、火灾报警、灭火以及发生火灾时用于人员疏散的火灾自动报警系统、自动灭火系统、消火栓系统、防烟排烟系统以及应急广播和应急照明、防火分隔设施、安全疏散设施等固定消防系统和设备。

- **Fire product /ˈfaɪə - ˈprɒdʌkt/:** products specially used for fire prevention, fire fighting, rescue, fire protection, refuge and escape

消防产品： 专门用于火灾预防、灭火救援和火灾防护、避难、逃生的产品。

- **Fixed extinguishing system /fɪkst - iksˈtɪŋgwiʃɪŋ - ˈsɪstəm/:** a fire extinguishing system that is fixedly installed in buildings, structures or facilities, etc., and consists of a fire extinguishing agent supply source, pipelines, spray devices and control devices

  固定灭火系统： 安装在建筑物、构筑物或设施等，由灭火剂供应源、管路、喷放器件和控制装置等组成的灭火系统。

- **Fire separation /ˈfaɪə - sepəˈreɪʃ(ə)n/:** to use building components with certain fire resistance to separate the internal space of the building and limit the fire to the fire area within a certain period of time

  防火分隔： 用具有一定耐火性能的建筑构件将建筑物内部空间加以分隔，目的是在一定时间内限制起火区的火灾。

- **Fire compartment /ˈfaɪə - kəmˈpɑːtm(ə)nt/:** the interior of the building separated by fire walls, fire-resistant floor slabs and other fire separation facilities, which can prevent fire from spreading to the rest of the same building within a certain period of time

  防火分区：在建筑内部采用防火墙、耐火楼板及其他防火分隔设施进行分隔，能在一定时间内防止火灾向同一建筑的其余地方蔓延。

- **Fire separation distance /ˈfaɪə - sepəˈreɪʃ(ə)n - ˈdɪst(ə)ns /:** the distance between adjacent buildings to prevent radiant heat from burning buildings to igniting adjacent buildings within a certain period of time, and to facilitate fire fighting

  防火间距：在一定时间内防止着火建筑的辐射热引燃相邻建筑，且便于消防扑救的间隔距离。

- **Fire wall /ˈfaɪə - wɔːl/:** non-combustible solid walls with duration of fire resistance not less than 3 hours to prevent fire from spreading to adjacent buildings or adjacent horizontal fire compartments

防火墙：防止火灾蔓延至相邻建筑或相邻水平防火分区且耐火极限不低于 3 小时的不燃性实体墙。

- **Fire bund /ˈfaɪə - bʌnd/:** a physical dam set up on the ground around the liquid storage tank to contain the leaked or spilled combustible liquid

防火堤：为容纳泄漏或溢出的可燃烧的液体，在液体储罐周围地面上设置的实体堤坝。

- **Fire resistance classification /ˈfaɪə - rɪˈzɪst(ə)ns - ˌklæsɪfɪˈkeɪʃ(ə)n/:** a classification of the overall fire resistance performance of buildings according to the different duration of fire resistance of various components such as walls, columns, beams, floor slabs, suspended ceilings, etc.

耐火等级：根据建筑中墙、柱、梁、楼板、吊顶等各类构件不同的耐火极限，对建筑物等整体耐火性能进行的等级划分。

- **Fire stability /ˈfaɪə - stəˈbɪlɪtɪ/:** under the conditions of standard fire test, the ability of load-bearing building components to resist collapse within a certain period of time

  耐火稳定性： 在标准耐火试验条件下，承重建筑构件在一定时间内抵抗坍塌的能力。

- **Fire integrity /ˈfaɪə - ɪnˈtegrɪtɪ/:** under the conditions of the standard fire resistance test, the ability to prevent flames and smoke from penetrating or flames appearing on the back fire side within a certain time when one side of a building partition is exposed to fire

  耐火完整性： 在标准耐火试验条件下，当建筑分隔构件一面受火时，在一定时间内防止火焰和烟气穿透到另一面的能力。

- **Fire insulation /ˈfaɪə - ɪnsjʊˈleɪʃ(ə)n/:** under the conditions of the standard fire resistance test, the ability to prevent the temperature of its backside surface from exceeding the specified value within a certain period of time when one side of the building partition is exposed to fire

耐火隔热性：在标准耐火试验条件下，当建筑分隔构件一面受火时，在一定时间内防止其背面温度超过规定值的能力。

- **Fire station /ˈfaɪə - ˈsteɪʃ(ə)n/:** barracks of the fire brigade, i.e., the most basic combat units for fire fighting and rescue, constructed and equipped with personnel, fire fighting equipment, training facilities, etc., in accordance with fire codes

  消防站：消防队的驻地，按照标准建设并配备人员、消防装备、训练设施等，是扑救火灾、抢险救援最基本的战斗单位。

- **Fire communication and command system /ˈfaɪə - kəmjuːnɪˈkeɪʃ(ə)n - ənd - kəˈmɑːnd - ˈsɪstəm/:** a communication and command system composed of networks, equipment and software, covering a certain area (such as a province or a city), connecting the units of the (mobile) fire command center, fire station, disaster relief and other links, with the function of fire alarm acceptance, communication dispatching, auxiliary decision-making, command and other functions

消防通信指挥系统：覆盖某一区域（如省、市），联通该区域的消防通信指挥中心、移动消防通信指挥中心、消防站、救灾相关单位等环节，具有火警受理、通信调度、辅助决策指挥和消防业务管理等功能的网络和设备及其软件组成的通信指挥系统。

- **Fire lookout tower /ˈfaɪə - ˈlʊkaʊt - ˈtaʊə/:** an observation facility in a certain height, which can be used to detect fires in time, send out fire alarms as soon as possible, and observe and report the fire situation

火警瞭望台：有一定高度的瞭望设施，利用它能及时发现火灾，及早发出火灾警报，并能观察与通报火场情况。

- **Fire lane /ˈfaɪə - leɪn/:** the passage of the building which meets the requirements of fire truck's access and operation, and is used exclusively by the fire brigade in emergency

  消防车通道：满足消防车通行和作业等要求，在紧急情况下供消防队专用，使消防员和消防车等装备能到达或进入建筑物的通道。

- **Fire equipment in building /ˈfaɪə - ɪˈkwɪpm(ə)nt - ɪn - ˈbɪldɪŋ/:** the general term for facilities installed in buildings and structures for fire alarm, fire fighting and rescue, personnel evacuation, and fire separation

  建筑消防设施：建筑物、构筑物中设置的用于火灾报警、灭火救援、人员疏散、防火分隔等设施的总称。

- **Fire detection and alarm system /ˈfaɪə - dɪˈtekʃ(ə)n - əˈlɑːm - ˈsɪstəm/:** a system that can detect early fire, send fire alarm and control signals to various fire fighting equipment to complete various fire fighting functions. It is generally composed of fire triggering devices, fire alarm devices, fire alarm controllers, and fire fighting linkage control systems.

火灾自动报警系统：能进行火灾早期探测、发出火灾报警信号，并向各类消防设备发出控制信号完成各项消防功能的系统，一般由火灾触发器件、火灾警报装置、火灾报警控制器、消防联动控制系统等组成。

- **Fire danger /ˈfaɪə - ˈdeɪn(d)ʒə/:** a collective term for fire hazard and fire risk

  火灾危险：火灾危害和火灾风险的统称。

- **Fire research /ˈfaɪə - rɪˈsɜːtʃ/:** an exploration of the mechanism, laws, characteristics, phenomena, effects and processes of fire

  火灾研究：针对火灾机理、规律、特点、现象、影响和过程的探求。

- **Fire resistance /ˈfaɪə - rɪˈzɪst(ə)ns/:** the ability of building components, fittings or structures to meet the stability, integrity and/or thermal insulation of the standard fire test within a certain period of time

耐火性能： 建筑构件、配件或结构在一定时间内满足标准耐火试验的稳定性、完整性和（或）隔热性的能力。

- **Fire retardant treatment /ˈfaɪə - rɪˈtɑːd(ə)nt - ˈtriːtm(ə)nt /:** the process used to improve the flame retardancy of materials

  阻燃处理： 用以提高材料的阻燃性。

- **Fire prevention /ˈfaɪə - prɪˈvenʃn/:** to take measures to prevent the occurrence of fire or limit the activities and processes it affects

  防火： 采取措施防止火灾发生，避免影响日常活动。

- **Fire point /ˈfaɪə - pɒɪnt/:** under the specified test conditions, the substance ignites on the surface under the action of an external ignition source and continues to combust the minimum temperature required for a given time

  燃点： 在规定的试验条件下，物质在外部引火源作用下起火并持续燃烧一定时间所需的最低温度。

- **Fire ground communication /ˈfaɪə - graʊnd - kəmjuːnɪˈkeɪʃ(ə)n/:** information transmission activities carried out on the fire ground in order to ensure the smooth progress of fire fighting and rescue operations
  火场通信： 为保证灭火救援行动的顺利进行，在火场上进行的信息传递活动。

- **Fire lookout /ˈfaɪə-ˈlʊkaʊt/:** the fire surveillance activities conducted by the watchman in the fire-lookout tower or other high locations in the jurisdiction
  火警瞭望： 瞭望员在火警瞭望台或其他高处对本辖区进行火情监视的活动。

- **Fire alarm /ˈfaɪə-əˈlɑːm/:** the emergency situations such as a fire
  火警： 发生火灾等紧急情况。

- **Fire telephone /ˈfaɪən - ˈtelɪfəʊn/:** the telephones specially used for report of fire alarm

火警电话：专门用于火灾报警的电话。

- **Fire telephone line /ˈfaɪə - ˈtelɪfəʊn - laɪn/:** the telephone line dedicated to report of fire alarm

  火警电话专钱：专门用于报告火灾的电话线路。

- **Fire risk management /ˈfaɪə - rɪsk - ˈmænɪdʒm(ə)nt/:** the processes, procedures and supporting cultural background required to obtain the expected fire risk standards

  Note: Fire risk management consists of fire-risk assessment, fire-risk treatment, fire-risk acceptance and fire-risk communication.

  火灾风险管理：获得预期的火灾风险标准所需的过程、程序和文化背景。

  注：火灾风险管理由火灾风险评估、火灾风险处置、火灾风险接受和火灾风险沟通组成。

- **Fire risk assessment /ˈfaɪə - rɪsk - əˈsesmənt/:** the process of evaluating the fire risk with the specified acceptable fire risk

火灾风险评估：用规定的可接受火灾风险对火灾风险进行评估的过程。

- **Fire risk acceptance** /ˈfaɪə - rɪsk - əkˈsept(ə)ns/: to decide whether to accept a fire-risk level according to the acceptance criteria

  火灾风险接受：根据可接受标准决定是否接受火灾风险水平。

- **Fire risk communication** /ˈfaɪə - rɪsk - kəmjuːnɪˈkeɪʃ(ə)n/: the act of communicating or sharing information on fire risk by risk related parties

  火灾风险沟通：风险相关方就火灾风险的信息进行交流或共享的行为。

- **Fire risk evaluation** /ˈfaɪə - rɪsk - ɪˌvæljʊˈeɪʃən/: to compare the estimated risk based on the fire-risk analysis with the acceptable risk based on the prescribed acceptance criteria

火灾风险评价：将基于火灾风险分析所估计的风险与基于可接受标准的风险进行对比。

- **Fire behaviour /ˈfaɪə - bɪˈheɪvjə/:** the changes in the physical and/or chemical properties of articles and (or) structures exposed to fire

  火灾特性： 物品和（或）构筑物暴露于火灾中所发生的物理和（或）化学变化。

- **Fire modeling /ˈfaɪə - ˈmɔdlɪŋ/:** to use the fire model to quantitatively describe the dynamic law of fire development

  火灾模化： 用火灾模型来定量地描述火灾发展的动态规律。

- **Fire test model /ˈfaɪə - test - mɔdl/:** a laboratory method used to describe a specific stage of a fire, including equipment, environment and test procedures

  火灾试验模型： 用于描述火灾特定阶段的试验室方法，包括设备、环境及试验程序。

- **Fire water supply facilities /ˈfaɪə - ˈwɔːtə - səˈplaɪ - fəˈsɪlətis/:** the artificial and natural water sources for fire fighting and rescue

  消防供水设施: 供灭火救援用的人工水源和天然水源。

- **Fire model /ˈfaɪə - ˈmɔdəl/:** an athematical expressions used to study and predict the development of fires

  火灾模型: 用于研究和预测火灾发展的数学表达式。

- **Fire scenario /ˈfaɪə - sɪˈnɑːrɪəʊ/:** a qualitative description of the entire process of a fire development, which identifies key events that reflect the characteristics of the fire and distinguish it from other possible fires

  Note: Fire scenarios usually define ignition, fire growth stage, full development stage and decline stage, as well as various systems and environmental conditions that affect the development of the fire. Regardless of whether the deterministic analysis or risk assessment is expected, it is an important step to identify the potential fire scenarios.

  火灾场景: 对整个火灾发生过程的定性描述, 该描述反映该次火灾特征并区别于其他潜在火灾的关键事件。

注：火灾场景通常要定义引燃、火灾增长阶段、完全发展阶段和衰退阶段，以及影响火灾发展过程的各种系统和环境条件。无论确定性分析或风险评估是否达到预期，确定预期的火灾场景都是重要的一步。

- **Fire lift /ˈfaɪə - lɪft/:** a set in the fire-resistant enclosed structure of the building, it has the front room, backup power supply and other fire protection, control and signal functions. It can be used by ordinary passengers under normal conditions and used exclusively by firefighters in the event of a fire in the building.

  消防电梯：建在建筑的耐火封闭结构内，具有前室、备用电源以及其他防火保护、控制和信号等功能，在正常情况下可为普通乘客使用，在建筑发生火灾时能专供消防员使用的电梯。

- **Fire plume /ˈfaɪə - pluːm/:** the upward turbulent flow formed by the buoyancy generated by combustion usually includes the lower combustion zone

火羽流：由燃烧产生的浮力形成的向上湍流，通常包括下部的燃烧区域。

- **Fire whirl** /ˈfaɪə - wɜːl/: the phenomenon of rapid rotation and flow of hot air caused by combustion

火旋风：因燃烧而引发的热空气快速旋转流动的现象。

- **Fire line** /ˈfaɪə - laɪn/: the boundary formed by the flame forward when the fire spreads

火线：由火蔓延时的火焰前锋所构成的界线。

- **Fire ground network** /ˈfaɪə - graʊnd - ˈnetwɜːk/: the wireless fire communication network formed between the commander of the fire ground and the commanders of the fire squadrons and combat squad leaders participating in the battle, namely, the secondary network

火场指挥网：火场指挥员与参战各公安消防中队指挥员及战斗班长之间组成的消防无线通信网，即二级网。

- **Fire communication room /ˈfaɪə - kəmjuːnɪˈkeɪʃ(ə)n - ruːm/:** the working room in the fire station that accepts fire alarms or dispatching instructions

  消防通信室：消防站内受理火警或接受调度指令的办公室。

- **Fire command diagram /ˈfaɪə - kəˈmaːnd - ˈdaɪəgræm/:** a schematic diagram reflecting the commander's intention to organize fire fighting and rescue operations

  火场指挥图：反映指挥员组织灭火救援作战意图的示意图。

- **Fire fighting commander /ˈfaɪə - ˈfaɪtɪŋ - kəˈmaːndə/:** the personnel who issue fire fighting and rescue orders and organize fire fighting and rescue on the fire ground

  灭火指挥员：在火场上发布灭火救援命令和组织实施灭火救援的人员。

- **Fighting deployment /ˈfaɪtɪŋ - dɪˈpl ɔɪmənt/:** after the fire brigade arrives at the fire ground, following the instruction of the fire fighting commander, firefighters quickly enter the designated position and complete the offensive preparation for the combustion area and the surrounding areas that need to be protected

战斗展开：消防队到达火场后，根据灭火指挥员的战斗命令，迅速进入指定位置，对燃烧区及周围需要保护的区域完成灭火准备。

- **Fire area /ˈfaɪə - ˈeərɪə/:** the urban and rural areas protected by the fire station

  辖区：消防站负责保护的城乡区域。

- **Fire brigade /ˈfaɪə - brɪˈɡeɪd/:** a fire fighting organization that is established in accordance with the law or as needed, equipped with personnel and fire fighting equipment, responsible for fire fighting and emergency rescue

  消防队：依法或根据需要建立，配备人员和消防装备等，负责火灾扑救、应急救援等工作的消防组织。

- **Fire fighting technique training /ˈfaɪə - ˈfaɪtɪŋ - tekˈniːk - ˈtreɪnɪŋ/:** the activities that enable trainees to master fire fighting methods and the operating skills of various fire fighting equipment

  灭火技能训练： 使受训人员掌握灭火方法和各种消防器材、消防装备等操作技能的训练活动。

- **Fire fighting training /ˈfaɪə - ˈfaɪtɪŋ - ˈtreɪnɪŋ/:** the exercises organized to adapt firefighters to the fire fighting plan or to test the effect of the plan, including fire fighting theories and techniques, tactics, physical training and psychological adaptability on the fire ground

  灭火战斗训练： 为使消防员适应灭火预案或检验灭火预案效果而组织的演练活动，包括灭火理论教育、灭火技能、灭火战术、身体素质及火场上的心理适应能力的训练等。

- **Fire fighting tactical training /ˈfaɪə - ˈfaɪtɪŋ - ˈtæktɪk əl - ˈtreɪnɪŋ/:** the training activities to familiarize and master various fire fighting tactics

灭火战术训练：熟悉、掌握各种灭火方式的训练活动。

- **Fire training tower /ˈfaɪə - ˈtreɪnɪŋ - ˈtaʊə/:** the buildings (structures) for firefighters to train in physical fitness, climbing skills, and high-rise fire fighting and rescue

  消防训练塔：供消防员进行身体素质、登高技巧和高楼灭火救援等训练的塔式建筑物。

- **Fire communication /ˈfaɪə - kəmjuːnɪˈkeɪʃ(ə)n/:** a communication command system composed of management and other functional networks, equipment and software, covering a certain area (province, municipality, autonomous region), connecting the fire communication command center, mobile fire communication command center, fire station, disaster relief related units and other links in the area, with fire alarm acceptance, communication dispatch, auxiliary decision-making command, and macro-management of fire protection business

  消防通信：某一区域（省、市、自治区），联通该区域的消防通信指挥中心、移动消防通信指挥中心、消防站、救灾相关单位等环节，具有火警受理、通信调度、辅助决策指挥和消防业务宏观管理等功能的网络和设备及其软件组成的通信指挥系统。

- **Fire break /ˈfaɪə - breɪk/:** a clear space in the direction of fire spread which is formed by using methods such as launching an upwind fire, removing fuel, or dampening potential fire sources

 防火带： 利用点燃逆风火、撤除燃料或浇湿潜在火源等方法在火灾蔓延方向上形成空旷地带。

- **Fire ground guard /ˈfaɪə - graʊnd - gɑ:d/:** to control personnel and vehicles entering and leaving the fire ground with measures such as demarcation of warning areas and traffic control, in order to ensure the smooth progress of fire fighting and rescue operations and fire investigations

 火场警戒： 为保证灭火救援行动和火灾调查的顺利进行，采取划分警戒区域、交通管制等措施对进出火场的人员、车辆进行控制。

- **First aid /ˈfɜːst - eɪd/:** firefighters take a series of fast and simple medical treatment measures on the wounded at the scene of fire fighting and rescue accidents to save the lives of the wounded, prevent the deterioration of the injury, alleviate the pain, prevent complications, and quickly and properly send the wounded to the hospital for treatment

  现场急救：消防员在灭火救援事故现场对伤员采取一系列快速、简捷的医疗处理措施，以挽救伤员的生命，防止伤情恶化，减轻伤痛，预防并发症，并迅速妥善地把伤员送到医院救治。

- **Fit test /ˈfɪt - test/:** the periodic test of how well the facepiece of a SCBA fits a particular firefighter

  密封性测试：定期测试自给式呼吸面罩是否适合特定消防员。

  注：SCBA 全称为 Self-contained Breathing Apparatus。

- **Flow meter /ˈfləʊ - miːtə/**
  流量计

- **Foam /fəʊm/:** an extinguishing agent formed by mixing foam concentrate with water and aerating the solution for expansion
  泡沫：通过将泡沫浓缩物与水混合使溶液充气膨胀而形成的灭火剂。

- **Force /fɔːs/:** the physical strength of something that is shown as it hits something else
  力；力量：某物击中其他物体时所表现出的力。

- **Forward lay /ˈfɔːwəd - leɪ/:** procedure of stringing water supply hose from a water source toward a fire scene; compare with reverse lay
  正铺设：从水源向火场辅设供水水带的行为，区别于反铺设。

- **Friction loss /ˈfrɪkʃ(ə)n - lɒs/:** it is the reduction of flow in a fire hose caused by friction between the water and the lining of the hose. It depends primarily upon diameter, type and length of the hose, and the amount of water flowing through.

摩擦损失：水和软管内壁之间的摩擦导致消防水带流量减少，主要取决于软管的直径、类型和长度，以及单位时间内流过的水量。

Note: GPM stands for gallons per minute.

注：GPM 指每分钟的加仑数。

- **Fully involved /ˈfʊlɪ - ɪnˈvɒlvd/:** the term of size-up meaning fire, heat and smoke in a structure are so widespread that internal access must wait until fire streams can be applied

  充分燃烧：现场评估的术语，指火势、热量、烟气蔓延范围太大，使用消防水枪灭火之前不得入内。

- **Flying fire /ˈflaɪɪŋ - ˈfaɪə/:** sparks or fireballs moving in the air

  飞火：在空中运动着的火星或火团。

- **Flameless combustion /fleɪmles - kəmˈbʌstʃ(ə)n/:** the substance is in a solid state without burning of flame

  无焰燃烧： 物质处于固体状态而没有火焰的燃烧。

- **Flash /flæʃ/:** combustible liquid volatilized vapor is mixed with air to reach a certain concentration, or combustible solid is heated to a certain temperature, and it flashes and combusts when it encounters an open flame

  闪燃： 可燃性液体挥发的蒸气与空气混合达到一定浓度或者可燃性固体加热到一定温度后，遇明火发生一闪即燃的现象。

- **Flashover /ˈflæʃəʊvə(r)/:** in a certain space, the surface of all combustible materials is involved in the transient process of combustion

  轰燃： 某一空间内，所有可燃物的表面全部燃烧的瞬变过程。

- **Flash point /flæʃ - pɒɪnt/:** under the specified test conditions, the minimum temperature at which the vapor

produced on the surface of the combustible liquid or solid
will flash under the action of the test flame

闪点：在规定的试验条件下，可燃性液体或固体表面产生的蒸气在试验火焰作用下发生闪燃的最低温度。

- **flat /flæt/:** a set of rooms (living room, bedroom, kitchen, etc.) for living in, usually on one floor of a building

公寓房；单元房；套房（每套房子包括客厅、卧室、厨房等房间）

- **Flame /fleɪm/:** a phase combustion area of luminous gas

火焰：发光气体相燃区。

- **Flammability /ˌflæməˈbiləti/:** under the specified test conditions, the ability of the material to undergo continuous flaming

易燃性：规定的试验条件下，材料发生持续有焰燃烧的能力。

- **Flame retardance /fleɪm - riˈtɑ:dəns/:** the property of a material to delay ignition or to inhibit, slow down or stop flame propagation

  阻燃性： 材料延迟被引燃或材料抑制、减缓或终止火焰传播的特性。

- **Foam-powder extinguishing system /fəʊmˈpaʊdə - iksˈtiŋgwiʃiŋ - ˈsɪstəm/:** a foam and dry powder combined application fire extinguishing system that can supply foam and dry powder separately, simultaneously or sequentially

  泡沫 - 干粉联用灭火系统： 可单独、同时或按顺序分别供给泡沫和干粉的两者联合应用灭火系统。

- **Foam extinguishing system /fəʊm - iksˈtiŋgwiʃiŋ - ˈsɪstəm/:** a fire extinguishing system that mixes foam fire extinguishing agent with water in a certain proportion and produces fire extinguishing foam through foaming equipment

  泡沫灭火系统： 将泡沫灭火剂与水按一定比例混合，

经发泡设备产生灭火泡沫的灭火系统。

- **Forcible entry /ˈfɔːsɪb(ə)l - ˈentri/:** firefighters destruct and demolish partial or all building components or other objects, in order to forcibly enter the fire ground for reconnaissance, rescue life on the fire ground, smoke exhaust, interception of fire spread, and evacuation of personnel and materials

  火场破拆：为强行进入火场进行火场侦察、火场救生、排烟、阻截火灾蔓延以及疏散人员和物资等行动，消防员对建筑构件或其他物体进行局部或全部破坏和拆除的活动。

- **Friction /ˈfrɪkʃ(ə)n /:** the resistance of one surface to another surface or substance moving over or through it

  摩擦力：一个表面对另一个表面的阻力。

- **Full-time fire brigade /ˈfulˈtaim - ˈfaɪə - brɪˈgeɪd/:** a professional fire brigade established in accordance with national laws and regulations, affiliated to local governments or enterprises and institutions, and responsible for fire fighting and emergency rescue

专职消防队：依照国家法律法规建立、隶属于地方政府或企事业单位、承担火灾扑救和应急救援工作的专业消防。

- **Flame front /fleɪm - frʌnt/:** the outer edge interface of the gas phase combustion zone on the surface of the material

  火焰前锋：材料表面上气相燃烧区的外缘界面。

- **Flammable /ˈflæməb(ə)l/:** things catch fire and burn easily

  易燃物：容易着火和燃烧的物体（东西）。

- **Fulcrum /ˈfʊlkrəm/:** the point on which a lever turns, balances, or is supported in turning or lifting something

  （杠杆的）支点：杠杆转动或支撑杠杆的点。

- **Fuse /fjuːz/:** a small wire or device inside a piece of electrical equipment that breaks and stops the current if the flow of electricity is too strong

保险丝；熔断器：电气设备内部的小电线或装置，如果电流太强，会断开并停止流动。

G

- **Gas fire extinguishing system /gæs - ˈfaɪə - iksˈtiŋgwiʃiŋ - ˈsɪstəm/:** a fire extinguishing system with gas extinguishing agent as its fire extinguishing medium
  气体灭火系统：以气体灭火剂作为灭火介质的灭火系统。

- **GPM:** gallons per minute or how many gallons are being pumped out of a piece of equipment every minute
  GPM：每分钟加仑数或每分钟从一台设备中抽出多少加仑。

- **Gauge /geɪdʒ/:** an instrument for measuring the amount or level of something

  计量器，测量仪表：测量某物数量或其他水平的仪器。

- **Guide rope /ˈgaɪdlaɪn - rəup/:** a rope employed by firefighters in environment with reduced visibility, smoke-filled zones and complex hazardous areas, to assist in establishing entry and exit directions while offering external guidance and protection

  导向绳：消防员进入浓烟区等能见度较低的场所及复杂的危险区域时，帮助其掌握进退方向或外部保护的绳子。

H

- **Hallway /ˈhɔːlweɪ/:** a long passage in a building with doors into rooms on both sides of it

  过道，走廊：建筑物里的长通道，两边都有通往房间的门。

- **Hazard /ˈhæzəd/:** something which could be dangerous to someone

  危险：可能对人产生危害的物质。

- **HAZMAT:** hazardous materials, including solids, liquids, or gases that may cause injury, death, or damage if released or triggered

危险品：包括固体、液体或气体，一旦释放或触发可能会造成人员伤亡或损害。

- **Halocarbon fire extinguishing system/ˌheɪləʊˈkɑːbən - ˈfaɪə - iksˈtɪŋgwiʃɪŋ - ˈsɪstəm/:** a gas fire extinguishing system composed of halogenated alkane supply sources, nozzles and pipelines

  卤代烷灭火系统：由卤代烷供应源、喷嘴和管路等组成的气体灭火系统。

- **Heat /hiːt/:** the added energy that causes substances to rise in temperature

  热量：使物质温度升高的附加能量。

- **Highlight /ˈhaɪlaɪt/:** to center attention on something

  突出，强调

- **High-rise building /ˈhaiˈraiz - ˈbɪldɪŋ/:** any building taller than three or four stories, depending upon local usage, requiring firefighters to climb stairs or aerial ladders for access to upper floors

高层建筑：高于三层或四层的建筑，根据当地的使用情况，需要消防员爬楼梯或用云梯进入高楼层。

- **High-rise pack/ˈhaiˈraiz - pæk/:** a shoulder load of hose with a nozzle and other tools necessary to connect the hose to a standpipe

  高层建筑包：肩扛式水带，带有喷嘴和其他将水带连接到立管的必要工具。

- **High wind warning /hai - wind - ˈwɔːnɪŋ/:** a sound and light alarm signal sent automatically by the fire alarm dispatching station when the wind reaches the preset wind level

  大风报警信号：风力达到预置风级后，火警调度台自动发出的声光报警信号。

- **Hot zone /hɒt - zəʊn/:** a contaminated area of HAZMAT

incident that must be isolated; requires suitable protective equipment to enter and decontamination upon exit; minimum hot zone distance from unknown material with unknown release is 330 feet (United Nations Emergency Response Guidebook); surrounded by "warm zone" where decontamination takes place

热区：危险事故中必须隔离的污染区；进入时需要合适的保护设备，离开时需要洗消；热区与未知物质的最小距离是 330 英尺（联合国应急指南）；热区周围是进行洗消的"暖区"。

- **High pressure carbon dioxide fire extinguishing system** /haɪ - ˈpreʃə - ˈkɑːb(ə)n - daɪˈɒksaɪd - ˈfaɪə - iksˈtiŋgwiʃiŋ - ˈsɪstəm/: a carbon dioxide fire extinguishing system in which carbon dioxide fire extinguishing agent is stored at room temperature

高压二氧化碳灭火系统：二氧化碳灭火剂在常温下贮存的二氧化碳灭火系统。

- **Hose /həʊz/:** a flexible tube made of rubber, plastic or canvas and used for directing water onto fires, etc.

软管: 由橡胶、塑料或帆布制成的柔性管,用于将水引向火场等处。

- **Hose reel system /həʊz - riːl - ˈsɪstəm/:** a hose system with a manually operated spray gun mounted on a reel or guide rail

  软管卷盘系统: 装在卷盘上或导轨上的可人工操作喷枪的软管系统。

- **Heat release rate /hiːt - rɪˈliːs - reɪt/:** the heat released by the combustion of materials or components per unit time

  热释放速率: 材料或组件在单位时间内燃烧所释放的热量。

- **Hydrant /ˈhaɪdr(ə)nt/:** a water pipe (esp. in a street) with a nozzle to which a fire hose can be attached, for drawing water from a water-main to put out fires, etc.

  消防栓,灭火龙头: 带喷嘴的水管,可连接消防水带,用于从自来水管引水灭火等。

- **Hydraulic rescue tools /haɪˈdrɔːlɪk - ˈreskjuː - tuːlz/**
  液压救援工具

- **Hydrogen /ˈhaɪdrədʒ(ə)n/:** a chemical element; a gas that is the lightest of all the elements and that is normally colorless, odorless, and highly flammable
  氢：化学元素；所有化学元素中最轻的一种气体，无色无味，但易燃。

I

- **Identification /aɪˌdɛntɪfɪˈkeɪʃən /:** (1) the process of showing, proving or recognizing who or what somebody or something is; (2) the process of recognizing that something exists, or is important

  识别，身份证明（常略作ID）；关联，联系：（1）显示、证明或确认某人是谁或某物或是什么的过程；（2）认识到某物存在或重要的过程。

- **IDLH:** Immediately Dangerous to Life and Health; the concentration of an air pollutant in a hazardous environment reaches a level that is dangerous, e.g., lethal, permanently damaging health, or immediately incapacitating someone to escape

立即危及生命和健康浓度：指有害环境中空气污染物浓度达到某种危险水平，如致命，永久损害健康或使人立即丧失逃生能力。

- **Ignition /ɪɡˈnɪʃ(ə)n/:** the devices to ignite the fuel in an internal-combustion engine

  点火装置：点燃内燃机燃料的装置。

- **Ignition source /ɪɡˈnɪʃ(ə)n - sɔːs/:** the external heat source (energy) that causes the substance to combust

  引火源：使物质开始燃烧的外部热源（能源）。

- **Ignition temperature/ɪɡˈnɪʃ(ə)n - ˈtemp(ə)rətʃə/:** the lowest temperature at which the substance ignites under the specified test conditions

  Note: GB/T 5332 specifies the test method for the ignition temperature of flammable liquids and gases.

引燃温度：在规定的试验条件下，物质发生引燃时的最低温度。

注：GB/T 5332 规定了可燃液体和气体引燃温度的测试方法。

- **Ignition time/ɪɡˈnɪʃ(ə)n - taɪm/:** under the specified test conditions, the time from when the sample is exposed to the specified thermal radiation conditions to the time of causing sustained combustion

  引燃时间：在规定的试验条件下，物质从开始暴露于规定的热辐射条件至达到持续燃烧的时间。

- **Incident commander /ˈɪnsɪd(ə)nt - kəˈmɑːndə/:** the officer in charge of all activities at an incident

  事故指挥官：负责事故中所有活动的官员。

- **Incident safety officer/ˈɪnsɪd(ə)nt - ˈseɪftɪ - ˈɒfɪsə/:** the officer in charge of scene safety at an incident

  事故安全员：负责事故中的现场安全的官员。

- **Indirect attack /ɪndɪˈrekt - əˈtæk/:** a method of fire fighting in which water is pumped onto materials above or near the fire so that water splashes onto the fire, often used where a structure is unsafe to enter

  间接攻击：将水抽到火上方或附近的材料上，使水花飞溅到火上的灭火方法，通常用于进入有危险的建筑物。

- **Ingredient /ɪnˈgriːdɪənt/:** the things that are used to make something

  原料：用来制作某种物品的东西。

- **Initial attack /iˈnɪʃəl - əˈtæk/:** the first point of attack on a fire where hose lines or fuel separation are used to prevent further extension of the fire

  初期攻击：火灾的第一个攻击点，使用水管或燃料分离来控制火势进一步蔓延。

- **Insulator /ˈɪnsjʊleɪtə/:** a material or device used to prevent heat, electricity, or sound from escaping from something

绝缘体：隔热（或绝缘、隔音等的）材料（或装置）。

- **Insulation /ɪnsjʊˈleɪʃ(ə)n/:** a thick layer of a substance that keeps something warm, especially a building
  隔热材料：使某物，尤指建筑物保持温暖的一层厚厚的物质。

- **Insurance /ɪnˈʃʊər(ə)ns/:** do something against something unpleasant happening
  保险措施：避免不愉快的事情发生而采取的措施。

- **Interior attack /ɪnˈtɪ3ːrɪə - əˈtæk/:** sending a team of firefighters into the burning structure, in an attempt to extinguish a blaze inside the structure, minimizing property damage fire, smoke, and water. It requires a minimum of four fully-equipped firefighters: an entry team of at least two entering the structure and fighting the fire, and two standing by to rescue or relieve the entry team. If the entry team(s) cannot extinguish the blaze, an exterior attack may be implemented.

内攻：派遣一组消防员进入着火的结构内部，尝试从结构内部扑灭大火，尽量减少火、烟和水造成的财产损失。要求至少有四名装备齐全的消防员：一个至少由两人组成的内攻组进入建筑内灭火，另外两人待命，以救援或替换内攻组的人员。如果内攻组不能扑灭大火，则可能变为外攻模式。

- **Inert /ɪˈnɜːt/:** the inhibition of the environment's ability to maintain combustion or explosion

  Note: For example, inert gas is injected into a closed space or a limited space to repel the oxygen inside to prevent fire.

  惰化： 对环境维持燃烧或爆炸能力的抑制。

  注： 例如把惰性气体注入封闭空间或有限空间，排斥里面的氧气，防止发生火灾。

- **Incapacitation /ˌɪnkəˌpæsəˈteʃən/:** the organism loses the ability to escape due to exposure to toxic gas or fire effluent

  丧失能力： 由于暴露于有毒气体或火灾流出物而使生物体失去逃生的能力。

- **Inerting system /iˈnə:t - ˈsɪstəm/:** a system that introduces an appropriate concentration of inert gas to prevent combustible gas, vapor, dust from burning or exploding

  惰化系统: 引入适当浓度的惰性气体防止可燃的气体、蒸气、粉尘燃烧或爆炸的系统。

- **Irons /ˈaiənz/:** the flat head axe mated with the Halligan bar. Firefighters often call these as the Crossed Irons, or Married Irons, because the Halligan bar can be inserted into the axe head.

  多功能铁铤: 平斧(板斧)与哈利根撬棍组合破拆工具。消防员通常将其称为"交叉铁铤",或"铁铤双璧",因为哈勒根撬棍可以插入平斧的头部配套使用。

- **ISO Rating /ˈreɪtɪŋ/:** it stands for Insurance Services Office Public Protection Classification Rating, and a rating published by the Insurance Services Office. Insurance companies, in many states of the USA, use this number to determine homeowner insurance premiums. However, some insurance companies have now adopted an actual loss system based on per zip code in several states and no longer use the ISO (PPC) system

ISO 评级：（保险服务局公共保护分类评级）这是由保险服务局公布的评级系统。美国许多州的保险公司都使用这个数字来确定房主的保险费。但是，有些保险公司已经采用了以每个邮政编码为基础的实际损失系统，不再使用 ISO（PPC）系统。

J

- **Jaws of life /dʒɔːz - əv - laɪf/:** hydraulic spreaders used in extrication procedures, most commonly used, but not limited to, during motor vehicle accidents
救生钳：解救过程中最常用的液压扩张器，不仅限于在机动车事故中使用。

- **Joule /dʒuːl/:** a unit of energy or work
焦耳：能量或功的单位。

- **Jurisdictional coverage network /ˌdʒuərisˈdikʃənəl - ˈkʌv(ə)** rɪdʒ - ˈnetwɜːk/: the fire control wireless communication network formed between the fire control dispatching command center and the communication command fire truck, the communication room of each fire squadron, and the communication fire truck, which is the primary network

辖区覆盖网： 消防调度指挥中心与通信指挥消防车、各消防中队通信室及通信消防车之间组成的消防无线通信网，即一级网。

L

- **Ladder company /ˈlædə - ˈkʌmp(ə)nɪ/:** a group of fire fighters, officers and engineers that staff a truck whose primary duty is to supply ladders to a fire scene. In most fire departments, the ladder company is responsible for ventilation of a structure on fire.

  云梯车连: 负责云梯车的消防员、指挥员和工程师，其主要职责是向火灾现场提供梯子。大多数消防部门中，云梯车连负责起火建筑的通风。

- **Laminar /ˈlæmɪnə/:** arranged in, consisting of, or

resembling laminae

层流的，层状的：排列成层状、由层状组成或类似层
状（的形态）。

- **Lever /ˈliːvə/:** a long piece of wood, metal, etc., used for lifting or opening something by somebody placing one end of it under an object and pushing down on the other end
  杠杆：一块长木头、金属等，将其一端放在物体下，另一端向下推，用来抬起或打开某物。

- **Liquid /ˈlɪkwɪd/:** a substance that flows freely and is not a solid or a gas, e.g., water
  液体：一种自由流动的物质，不是固体或气体，例如水。

- **Liquefy /ˈlɪkwɪfaɪ/:** to become liquid; to make something liquid
  （使）液化：使某物变为液体。

- **Lobby** /ˈlɒbɪ/: a large area inside the entrance of a public building where people can meet and wait

  一层大厅，（公共建筑物进口处的）门厅，前厅，大厅

- **Level I, II, III incident:** a HazMat term commonly used to describe different levels of denoting the severity of the incident and the type of response that may be necessary, where Level III is the largest or most dangerous

  I 级、II 级、III 级事件：危险品术语，表示事件的严重程度和可能需要的反应类型，其中 III 级是最严重或最危险的。

- **Local terminal** /ˈləʊk(ə)l - ˈtɜːmɪn(ə)l/: a terminal directly connected to the host computer set in the fire command center

  本地终端：设置在消防指挥中心的直接与计算机主机相连的终端。

- **Local application of high/medium expansion foam extinguishing system** /ˈləʊk(ə)l - æplɪˈkeɪʃ(ə)n - əv - haɪ/

ˈmiːdɪəm - ɪkˈspænʃ(ə)n - fəʊm - iksˈtɪŋgwiʃɪŋ - ˈsɪstəm/: a foam extinguishing system in which a fixed or semi-fixed high or medium foam generating device sprays foam to the fire site directly or through a bubble guide tube

局部应用式高倍数、中倍数泡沫灭火系统：由固定式或半固定式高倍数或中倍数泡沫发生装置直接或通过导泡筒将泡沫喷放到火灾部位的泡沫灭火系统。

- **Low pressure carbon dioxide fire extinguishing system /** ləʊ - ˈpreʃə - ˈkɑːb(ə)n - daɪˈɒksaɪd - ˈfaɪə - iksˈtɪŋgwiʃɪŋ - ˈsɪstəm/: a carbon dioxide fire extinguishing system that the fire extinguishing agent is stored at a temperature of -20°C to -18°C.

低压二氧化碳灭火系统：二氧化碳灭火剂在–20℃~18℃的温度下贮存的二氧化碳灭火系统。

M

- **Main /meɪn/:** a principal pipe bringing water or gas, or a principal cable carrying electric current, from the source of supply into a building

  （自来水、煤气等的）总管道；干线

- **Mains /meɪnz/:** the source of water, gas or electricity supply to a building or area

  （供应建筑物或地方的）水源、煤气源、电源

- **Maintenance /ˈmeɪnt(ə)nəns/:** the act of keeping

something in good condition by checking or repairing it regularly

维护；保养：通过定期检查或修理使某物保持良好状态的行为。

- **Maltese Cross /mɔlˈtiz - krɒs/:** the emblem of the fire service is often referred to as a "Maltese Cross." But the actual origin of the current or common emblem in the USA remains uncertain. While it is true that the Knights Hospitalers (also known as Knights of St. John of Jerusalem), who were based in Malta 1530—1798, did wear a cross emblem and a version of that cross has been used as a fire service icon, it bears little resemblance to the current form in use in much of the United States. It is possible to accept that the current design is just a stylized artistic embellishment of the original form. The current design may also have been influenced by the design of the cross of Saint Florian.

  马耳他十字："美国消防部门的徽章通常被称为"马耳他十字"。但目前常见的徽章起源仍不确定。虽然医院骑士团（又称耶路撒冷圣约翰骑士团）确实佩戴过十字架标志，而且其中一个版本也被用作消防服务标志，但与美国大部分地区的标志几乎没有相似之处。目前的设计只是对原始形式的风格化的艺术润色，可能也受到了圣弗洛里安十字架设计的影响。

- **Malicious fire alarm /məˈlɪʃəs - ˈfaɪə - əˈlɑːm/:** deliberately issuing an alarm of fire knowing that there is no fire
  谎报火警： 明知没有火灾等情况发生而故意发出火灾警报。

- **Manufacture /mænjʊˈfæktʃə/:** to make something in a factory, usually in large quantities
  生产： 在工厂里制造产品，通常是大量生产。

- **Masonry /ˈmeɪs(ə)nrɪ/:** bricks or pieces of stone which have been stuck together with cement as part of a wall or building
  砖石建筑： 用水泥粘在一起的砖或石片构成部分墙体

或建筑。

- **Mass casualty incident (MCI) /mæs - ˈkæʒjʊəltɪ - ˈɪnsɪd(ə)nt/:** any incident that produces a large number of injured persons requiring emergency medical treatment and transportation to a medical facility. The exact number of injured persons that makes an incident "mass casualty" is defined by departmental procedures and may vary from area to area.

  大规模伤亡事件（MCI）： 产生大量受伤人员需要紧急医疗和运送到医疗机构的事件。"大规模伤亡"事件的伤员数量标准由部门程序规定，可能各地不同。

- **Mass burning rate /mæs - ˈbɜːnɪŋ - reɪt/:** under the specified test conditions, the mass loss caused by the combustion of the material per unit time

  质量燃烧速率： 在规定的试验条件下，材料在单位时间内燃烧造成的质量损失。

- **Mass /mæs/:** the quantity of material that something

82222

contains

质量：物体所含物质的量

- **Matter /ˈmætə/:** the physical part of the universe consisting of solids, liquids, and gases

物质：宇宙中由固体、液体和气体组成的物理部分。

- **Master stream /ˈmɑːstə - striːm/:** a large nozzle, either portable or fixed to a pumper, capable of throwing large amounts of water relatively long distances

主流：一个大型的喷嘴，可以是便携式的，也可以是固定在泵车上的，能够将大量的水射出相对较远的距离。

- **Means of egress /miːnz - əv - ˈiːgres/:** the way out of a building during an emergency; may be by door, window, hallway, or exterior fire escape. Local fire codes will often dictate the size, location and type according to the number of occupants and the type of occupancy.

出口方式：在紧急情况下离开建筑物的途径，可以通

110

过门、窗、走廊或外部消防通道。当地的消防法规通
常会根据居住者的数量和居住类型来决定其大小、位
置和类型。

- **Mercury /ˈmɜːkjuri/:** a silver-coloured liquid metal
  水银

- **Methane /ˈmiːθeɪn/:** a gas without colour or smell that
  burns easily and is used as fuel
  甲烷：一种无色无味的易燃气体，用作燃料。

- **Mobile terminal /ˈməʊbaɪl - ˈtɜːmɪn(ə)l/:** a terminal,
  which is on the communication command fire truck,
  transmits information through a wireless channel and is
  connected to the command center computer
  移动终端：在通信指挥消防车上，通过无线信道传输
  信息并与指挥中心计算机相连的终端。

- **Molecule /ˈmɒlɪkjuːl/:** the smallest particle of a substance
  that retains all the properties of the substance and is

composed of one or more atoms

分子：分子是保持物质化学性质的最小粒子，由一个或多个原子组成。

- **Momentum /məˈmentəm/:** the quantity of movement of a moving object, measured as its mass multiplied by its speed

  动量：运动物体在它运动方向上保持运动的趋势，是物质的质量和速度的乘积。

- **Mechanics /məˈkænɪks/:** a brach of physics that deals with energy and forces and their effect on bodies

  力学：物理学的一个分支，主要研究能量和力对物体的影响。

- **Mechanical pressurization /mɪˈkænɪk(ə)l - ˌprɛʃrɪˈzefən/:** use mechanical air supply to the stairwell, front room and other areas that need to be protected to form a positive pressure in the area to prevent smoke from entering

  机械加压送风：对楼梯间、前室及其他需要被保护的区域采用机械送风，使该区域形成正压，防止烟气进入的方式。

- **Mechanical smoke extraction /mɪˈkænɪk(ə)l - sməʊk - ɪkˈstrækʃ(ə)n/:** the smoke exhaust method that uses mechanical force to exhaust the smoke to the outside of the building

  机械排烟： 采用机械力将烟气排至建筑物外的排烟方式。

- **Minimum ignition time /ˈmɪnɪməm - ɪgˈnɪʃ(ə)n - taɪm/:** under the specified test conditions, the shortest time for a substance to ignite when exposed to heat radiation

  最小引燃时间： 在规定的试验条件下，物质暴露于热辐射条件下而发生燃烧的最短时间。

- **Miscibility /ˌmɪsəˈbɪləti/:** a property of substances that can be mixed together

  可混合性

- **Melt drip /melt - drɪp/:** the dripping when a substance burns or melts

  熔滴： 物质燃烧或熔融时的滴落物。

- **Melting behaviour /ˈmeltɪŋ - bɪˈheɪvjə/:** the physical phenomena such as shrinkage, dripping, and melting of substances occur when heated

  熔化行为： 物质受热发生皱缩、滴落、熔化等物理现象。

- **Mutual aid /ˈmjuːtʃʊəl - eɪd/:** an agreement between nearby fire companies to assist each other during emergencies by responding with available manpower and apparatus

  相互援助： 附近的几个消防连队之间的合作，在紧急情况下以现有的人力和设备协助对方。

- **Multiple alarms /ˈmʌltɪp(ə)l - əˈlɑːms/:** A request by an incident commander for additional personnel and apparatus. Each department will vary on the number of apparatus and personnel on each additional alarm.

多重警报：事故指挥官提出的增加人员和设备的要求。每个部门在每个额外警报的设备和人员的数量方面有所不同。

- **MVA:** Motor vehicle accident

  机动车事故

N

- **Natural smoke extraction /ˈnætʃ(ə)r(ə)l - sməʊk - ɪkˈstrækʃ(ə)n/:** the smoke exhaust method that uses the buoyancy of the hot smoke flow generated during a fire and the effect of external wind to exhaust the smoke to the outside through the external opening of a building

  自然排烟：一种排烟方式，利用火灾时产生的热烟气流的浮力和外部风力作用，通过建筑物的对外开口把烟气排至室外。

- **NFPA:** the National Fire Protection Association, a research group in the USA which sets a number of standards and best practices for fire fighting, equipment, and fire protection in the USA, which is also adopted in many other countries

  美国国家消防协会： 为美国的消防、设备和防火制定许多标准和最佳实践的研究机构，许多其他国家也采用了 NFPA 制定的标准。

- **NIMS:** the National Incident Management System, of the USA, a federally mandated program for the standardizing of command terminology and procedures. This standardizes communications between fire departments and other agencies. It is based upon simple terms that will be used nationwide. Currently, the federal government of the USA requires DHS and FEMA to organize training programs, which is in the process of standardizing terms and procedures under the NIMS.

  国家突发事件管理系统： 美国联邦政府授权的计划，旨在使指挥术语和程序标准化。该计划用美国范围内使用的基本术语规范了消防部门和其他机构之间的沟通。目前，美国联邦政府要求国土安全部和联邦应急管理局组织培训项目，将国家突发事件管理系统下的术语和程序进行规范统一。

- **NIOSH:** the National Institute for Occupational Safety and Health, an agency of the USA responsible for investigation of workplace deaths, including firefighters 国家职业安全健康研究所：负责调查包括消防员在内的工作场所亡故事件的美国机构。

O

- **Occupancy /ˈɒkjəpənsɪ/:** zoning and safety code term used to determine how a structure is permitted to be used and occupied, which in turn dictates the necessary safety structures and procedures

  占用：分区和安全代码术语，用于确定建筑物使用和占用的方式，规定必要的安全结构和程序。

- **Offensive attack /əˈfensɪv - əˈtæk/:** a method of fire fighting in which water or other extinguisher is taken directly to the site of the fire, as opposed to being pumped in that general direction from a safe distance

进攻式灭火: 直接将水或灭火器带到火点的灭火方法，而不是在安全距离内灭火。

- **Ohm /əʊm/:** a unit which is used to measure electrical resistance

  欧姆: 电阻单位。

- **Open automatic sprinkler system /ˈəʊpən - ˌɔːtəˈmætɪk - ˈsprɪŋklə - ˈsɪstəm/:** an automatic sprinkler system using open sprinkler head

  Note: Including deluge extinguishing system, drencher extinguishing system, water spray extinguishing system, etc.

  开式自动喷水灭火系统: 采用开式洒水喷头的自动喷水灭火系统。

  注: 包括雨淋灭火系统、水幕式灭火系统、水喷雾灭火系统等。

- **Operation /ˌɒpəˈreɪʃn/:** working; activity

行动：操作；活动。

- **Optical density of smoke /ˈɒptɪkl - ˈdensəti - əv - sməʊk/:** the degree of attenuation of a beam passing through a flue gas, which is expressed by the logarithm of the flue gas light blocking rate

Note 1: Smoke density is dimensionless.

Note 2: The flue gas light blocking rate refers to the ratio of the intensity of human light emitted by the intensity of light transmitted through flue gas under the specified test conditions, which is the reciprocal of the transmittance.

光学烟密度： 用烟气阻光率对数表示的光束穿过烟气的衰减程度。

注 1：烟密度是无量纲的。

注 2：烟气阻光率是指在特定的试验条件下，人体发出的光强度与透过烟气的光强度之比，是透射率的倒数。

- **Organization and command of fire fighting and rescue operation /ˌɔːgənaɪˈzeɪʃn - ənd - kəˈmɑːnd - əv - ˈfaɪəfaɪtɪŋ - ənd - ˈreskjuː - ˌɒpəˈreɪʃn/:** the special organization and leadership activities of commanding organizations and commanders with fire fighting and rescue command qualifications to deal with various disasters and accidents, which runs through the entire process from the dispatch of the firefighters to the restoration of combat readiness

灭火救援组织指挥：具备灭火救援指挥资格的指挥机关和指挥员对各类灾害事故进行处置的特殊的组织领导活动，贯穿于派遣消防员至恢复战备的全过程。

- **OSHA:** the Occupational Safety and Health Administration, the USA government agency concerned with regulating employee safety, particularly in hazardous occupations such as fire fighting

  OSHA：职业安全卫生管理局，美国政府机构，负责监管员工安全，特别是消防等危险职业。

- **Overhauling /ˌovəˈhɔliŋ/:** a late stage in fire-suppression process during which the burned area is carefully examined for remaining sources of heat that may re-kindle the fire. It often coincides with salvage operations to prevent further loss to structure or its contents, as well as fire-cause determination and preservation of evidence.

清理火场：灭火过程的后期阶段，在此期间仔细检查燃烧区域是否存在可能重新引起火灾的剩余热源。通常与救援工作同时进行，以防止结构或其物品的进一步损失，以及确定火灾原因和保存证据。

- **Oxidizer /ˈɔksidaizə/:** an oxidant, especially a substance that combines with the faul in a rocket engine; an oxidizing agent, especially, one used to support the combustion of a rocket propellant

  氧化剂：尤指用于火箭推进剂中的氧化物；用来帮助火箭燃烧的氧化剂。

- **Oxidant /ˈɒksɪd(ə)nt/:** a substance that makes another substance combine with oxygen

  氧化物：使另一种物质与氧结合的物质。

- **Oxygen index /ˈɒksɪdʒən - ˈɪndeks/:** the minimum oxygen concentration required for flame-burning of the material in a nitrogen and oxygen mixture under the specified test conditions

Note: the unit of oxygen index is "%".

氧指数：在特定的试验条件下，材料在氮氧混合气中进行有焰燃烧所需的最低氧浓度。

注：氧指数的单位为"%"。

P

- **Panic /ˈpænɪk/ :** a great fear

  恐慌

- **Particle/ˈpɑːtɪkl/:** a very small piece of something

  微粒

- **Parapet /ˈpærəpɪt/:** a low protective wall along the edge of a balcony, bridge, roof, etc.; a protective bank of earth, stones, etc., along the front edge of a trench in a war

  矮护墙，女儿墙；胸墙：沿阳台、桥梁、屋顶等的边缘修建的低矮防护墙；（在战争中）沿战壕前沿用土、石头等修建的防护堤岸。

- **Pascal /ˈpæskəl/:** the standard unit for measuring pressure

  帕（斯卡）： 标准压强单位。

- **Personnel Accountability Report (PAR) /ˌpɜːsəˈnel - əˌkaʊntəˈbɪləti - rɪˈpɔːt/:** an end-result of personnel accountability system. Best report is all hands, while worse is squad missing. You will often hear command ask for a "PAR" when something has changed on the fire ground. Often the reply will be something like, "Engine 4, PAR." or "Engine 4 has PAR."

  火场责任制报告（PAR）： 火场责任制度的最终结果。最佳报告是"所有人一切正常"，较差的是"小队失踪了"。当火情有变时，经常会听到指挥部询问"PAR"。回复通常是"Engine 4, PAR"或"Engine 4 has PAR"。

- **Personnel accountability system /ˌpɜːsəˈnel -əˌkaʊntəˈbɪləti - ˈsɪstəm/:** tag, passport, or other system for identification and tracking of personnel at an incident, especially those entering and leaving an IDLH area; intended to permit rapid determination of those who may be at risk or lost during sudden changes at the scene

  火场责任系统：标签、通行证或其他系统，用于识别和跟踪事故中的人员，特别是进出威胁生命和健康(IDLH)区域的人员，旨在快速确认在现场异常情况下有危险或迷失方向的人员。

- **Personnel density /ˌpɜːsəˈnel - ˈdensəti/(theory):** the number of personnel per unit building area, which is used to calculate the number of emergency exits and their width

  人员密度（理论）：单位建筑面积上的人员数目，用于计算安全出口数量和出口宽度。

- **Piston effect /ˈpɪstən - ɪˈfekt/:** the forced air flow inside a tunnel caused by moving vehicles

活塞效应：指在隧道内由于车辆的移动而产生的强制气流。

- **Pivot /ˈpɪvət/:** the central point, pin or column on which something turns or balances

  支点；枢轴：物体旋转或保持平衡的中心点。

- **Plug /plʌg/:** an old term for a fire hydrant. This survives from the days when water mains actually had holes in the tops that were plugged. Many firefighters would like to keep this word while many others think it should be replaced with the accurate term, "hydrant."

  消防栓：fire hydrant 的旧表述。这个词是在水管顶部有孔并被堵塞的年代遗留下来的。许多消防员希望保留这个词，还有一些人认为应该用准确的术语"hydrant"来代替。

- **Pneumatic equipment /njuːˈmætɪk - ɪˈkwɪpmənt/**

  气动设备

- **Portable fire extinguisher** /ˈpɔːtəbl - ˈfaɪə(r) - ɪkˈstɪŋgwɪʃə/
  手提式灭火器

- **Power** /ˈpaʊə(r)/: the energy that can be collected and used to operate a machine
  能：用于操作机器的能量。

- **Power tools** /ˈpaʊə(r) - tuːlz/
  电动工具

- **Powder extinguishing system** /ˈpaʊdə(r) - ɪksˈtɪŋgwɪʃɪŋ - ˈsɪstəm/: a fire extinguishing system composed of dry powder storage containers, drive components, conveying pipes, spray components, detection and control devices, etc.
  干粉灭火系统：由干粉贮存容器、驱动组件、输送管道、喷放组件、探测和控制器件等组成的灭火系统。

- **Pool fire** /puːl - ˈfaɪə(r)/: a fire that occurs in a pool containing flammable, combustible liquids or dissolved solids

油池火：在含有易燃、可燃液体或溶解固体池内发生的火灾。

- **Pre-arrival instructions /pri:əˈraɪvl - ɪnˈstrʌkʃnz/:** the directions given by a dispatcher to a caller until emergency units can arrive

  到达前指示：在紧急救援人员到达之前，调度员向呼叫者提供的指示。

- **Pre-action automatic sprinkler system /pri:ˈækʃ(ə)n - ˌɔ:təˈmætɪk - ˈsprɪŋklə - ˈsɪstəm/:** in the quasi-working state, the water distribution pipeline is not filled with water. The automatic fire alarm system and closed sprinklers are used as detection elements. After the rain alarm valve or pre-action alarm valve group is automatically opened, it is converted to a wet automatic sprinkler system of a sealed automatic sprinkler system.

  预作用自动喷水灭火系统：在准工作状态下，配水管道没有充满水，由自动火灾报警系统和封闭式洒水喷头作为检测元件，雨淋报警阀或预作用报警阀组自动开启后，将其转换为湿式自动喷水灭火系统的封闭式自动喷水灭火系统。

- **Precaution /prɪˈkɔːʃn/:** an action that is intended to prevent something dangerous or unpleasant from happening
  防范行动：防止发生危险或意外的行动。

- **Pre-determined fire fighting plan / priːdɪˈtɜːmɪnd - ˈfaɪəfaɪtɪŋ - plæn/:** fire fighting operation plan, a fire fighting and rescue operation document made in advance according to the expectation of the fire target, the dispatchable rescue force and the disaster situation
  灭火预案：灭火作战计划。根据火灾的预测以及可调度的救援力量和灾情分析，提前制定的灭火救援行动。

- **Pre-fire, pre-incident planning /priːˈfaɪə(r) - priːˈɪnsɪdənt - ˈplænɪŋ/:** information collected by fire prevention officers to assist in identifying hazards and the equipment, supplies, personnel, skills, and procedures needed to deal with a potential incident

火灾前、事故前规划：防火员收集的信息以帮助识别危险以及提供潜在事故所需的设备、物资、人员、装备和程序。

- **Pre-mixed flame /ˌpriːˈmikst - fleɪm/:** a flame produced when a fuel is pre-mixed with an oxidizer and then ignited

  预混火焰：燃料与氧化剂混合后点燃所产生的火焰。

- **Pre-planning /priːˈplænɪŋ/:** a fire protection strategy involving visits to potentially hazardous occupancies for inspection, follow-up analysis and recommendations for actions to be taken in case of specific incidents. Not to be confused with post-planning.

  预案：消防策略，包括对潜在危险场所进行检查、跟踪分析并对发生具体事故时应采取的行动提出建议，不要与后期规划相混淆。

- **Pressure/ˈpreʃə(r)/:** the force produced by a particular amount of gas or liquid in a confined space or container

  压力：用力按压物体时产生的力。

- **Primer /ˈpraɪmə/:** a small explosive device that is used to set off a larger explosion

  （爆炸物的）引信、底火：一种小型爆炸装置，用来引发更大的爆炸。

- **Probie /ˈprəubi/:** (also rookie) a new firefighter on employment probation (a period of time during which his or her skills are improved, honed, tested, and evaluated)

  新手：（在美国也称作 "rookie"）试用期内的新消防员（在此期间他或她的技能得到提高、磨练、测试和评估）。

- **Product of combustion /ˈprɒdʌkt - əv - kəmˈbʌstʃ(ə)n/:** all substances produced by combustion or pyrolysis

  燃烧产物：由燃烧或热解作用而产生的全部物质。

- **Professional firefighter /prəˈfeʃənl - ˈfaɪəfaɪtə(r)/:** all firefighters are classified as "professionals" by both the International Association of Fire Chiefs (IAFC) and the International Association of Fire Fighters (IAFF trade union). All firefighters are required by most state laws of the USA and general practice to meet the same training and equipment standards, take the same examinations for promotion and perform the same work under the same hazards. There are two accepted categories of professional firefighters: volunteer firefighters who may or may not receive pay for services and career firefighters whose primary employment and source of earned income is in the fire service.

  专业消防员：所有消防员都被国际消防长官协会（IAFC）和国际消防员协会（IAFF 工会）列为"专业人员"。美国大多数州的法律和惯例要求所有消防员必须达到统一培训标准和设备使用标准，参加统一的晋升考试并在相同的危险下执行相同任务。专业消防员有两种公认的类别：志愿消防员和职业消防员。

志愿消防员不一定获得服务报酬，职业消防员的主要
工作和收入来源是消防部门。

- **Property /ˈprɒpəti/:** the ways in which it behaves in
  particular conditions
  (物质、物体的) 特性: 物质在特定条件下发生的反应。

- **Public fire facilities /ˈpʌblɪk - ˈfaɪə(r) - fəˈsɪlətiz/:** the
  necessary public facilities to ensure fire fighting safety,
  which usually include fire stations, fire communication
  command system, fire lookout tower, fire water facilities
  and fire engine access
  公共消防设施: 保障消防安全的必要公共设施，通常
  包括消防站、消防通信指挥系统、火警瞭望台、消防
  供水设施和消防通道。

- **Pump /pʌmp/:** machine or device for forcing liquid, gas
  or air into, out of or through something, e.g., water from
  a well, petrol from a storage tank, air into a tyre or oil
  through a pipeline

泵，抽水机；打气筒：使液体、气体或空气进入、排出或通过某物（如：从井中抽水、从储油罐中抽汽油、往轮胎中打气、或往管线中泵油）的机器或装置。

- **Pumper /ˈpʌmpə/:** a fire truck with a water tank

  泵车：配备水箱的消防车。

- **Pump operator /pʌmp - ˈɒpəreɪtə(r)/** (also a chauffeur): a person who is responsible for operating the pumper and typically for driving the pumper to the scene of the accident

  水泵操作员、技术员（也称作司机）：负责操作消防车上的水泵，并且将消防车驾驶到事故现场的人员。

- **Pyrolysis /paɪˈrɒlɪsɪs/:** the substance undergoes irreversible chemical decomposition without oxidation due to temperature increase

  热解：物质经历不可逆的化学分解，且不会因温度升高而氧化。

- **Pyrophoric material /ˌpaɪrə(ʊ)ˈfɒrɪk - məˈtɪəriəl/:** a substance that can combust on its own when it contacts with air

  自燃物：与空气接触即能自行燃烧的物质。

R

- **Radiation /reɪdɪˈeɪʃ(ə)n/:** the emission of energy

  辐射：能量的释放。

- **Radiator /ˈreɪdɪeɪtə/:** a hollow metal device for heating rooms

  散热器，散热片：用于加热房间的中空金属装置。

- **Rapid Intervention Crew/Group/Team (RIC, RIG, or RIT) /ˈræpɪd - ˌɪntəˈvenʃn - kruː/ /gruːp/ /tiːm/:** this is a standby crew whose purpose is to go in for the rescue of firefighters in trouble. While all of these versions of the names for a firefighter rescue crew either have been used or continue to be used in several areas, the National Incident Management System (NIMS) of the USA has adopted the term Rapid Intervention Crew/Company ("RIC") to be the standard in the Incident Command System (ICS).

快速干预小组 / 组 / 队（RIC、RIG 或 RIT）：待命小组，其职责是救援遇险。各版本的名称都已在美国多个领域使用或继续使用，但美国国家事故管理系统 (NIMS) 已采用 RIC 作为事件指挥系统 (ICS) 中的标准术语。

- **Reaction to fire /riˈækʃn - tuː - ˈfaɪə(r)/:** the reaction of a material or product in contact with fire under specified test conditions

  对火反应： 在规定的试验条件下，材料或制品遇火所产生的反应。

- **Receipt of fire alarm /rɪˈsiːt - əv - ˈfaɪə(r) - əˈlɑːm/:** the activity of a fire brigade to receive information about the occurrence of a fire

接警：消防队接收发生火灾信息的活动。

- **Reconnaissance on the fire ground /rɪˈkɒnɪs(ə) ns - ɒn - ðə - ˈfaɪə(r) - graʊnd/:** the activity that is implemented to grasp the situations of the fire ground by means of observation, inquiry, and detection after the fire brigade arrives at the fire ground

  火场侦察：消防队到达火场后，通过观察、调查、侦测等方法全面掌握火场情况的活动。

- **Reflash /riːˈflæʃ/ or rekindle /riːˈkɪndl/:** a situation in which a fire thought to be extinguished, resumes burning

  再燃，复燃：已熄灭的火又重新燃烧起来。

- **Reflection /rɪˈflekʃn/:** the action or process of sending back light, heat, sound, etc., from a surface

  （声、光、热等的）反射：从表面发回光、热、声等的动作或过程。

- **Refuge floor** /ˈrefjuːdʒ - flɔː/ˌ; **Refuge room** /ˈrefjuːdʒ - ruːm/: a floor or room in a building used for temporary shelter from fire and its fumes in case of fire

  避难层；避难间：为应对火灾所提供的临时避火和烟雾的楼层或房间。

- **Relay** /ˈriːleɪ/: a fresh set of people or animals taking the place of others who have finished a period of work

  接班接力：新的人或动物代替长时间工作的其他人或动物。

- **Relay transmission** /ˈriːleɪ - trænzˈmɪʃn/: the live signal of the fire site at the fire ground TV transmitting location is transmitted to the receiving site through one or more wireless relay equipment

  中继传输：火灾现场电视发射位置的现场信号通过一个或多个无线中继设备传输到接收现场。

- **Remote terminal** /rɪˈməʊt - ˈtɜːmɪnl/: a terminal, which is located in the fire squadron, uses for the dedicated fire dispatch line to transmit information and connects to the command center computer through a modem

远程终端：消防中队的终端使用消防调度专线传输信息，并通过调制解调器连接到指挥中心计算机。

- **Report of fire alarm** /rɪˈpɔːt - əv - ˈfaɪə(r) - əˈlɑːm/: to report the fire to the fire brigade
  火灾报警：向消防队报告火灾。

- **Representative fire scenario** /ˌreprɪˈzentətɪv - ˈfaɪə(r) - səˈnɑːriəʊ/: a representative fire scenario selected from a fire scene group, assuming that the result can provide a reasonable estimation of the average result of the fire scene group
  典型火灾场景：从火灾现场组中选择具有代表性的火灾场景，且结果可对火灾现场组的平均结果提供合理估计。

- **Request for reinforcement** /rɪˈkwest - fɔː - ˌriːɪnˈfɔːsmənt/; **assistance message** /əˈsɪstəns - ˈmesɪdʒ/: information issued by the fire ground or other emergency scenes requesting reinforcement of fire trucks, equipment and personnel

请求增援；援助消息：由火场或其他紧急事件现场发出的要求增援消防车、设备和人员的信息。

- **Residential** /ˌrezɪˈdenʃl/: a residential area contains houses rather than offices or factories

  住宅的：只有住宅，不含办公室或工厂的居住地。

- **Rescue company** /ˈreskjuː - ˈkʌmpəni/: a squad of firefighters is trained and equipped to enter adverse conditions and rescue victims of an incident, and it is often delegated to a truck company

  救援车连：训练有素、装备精良的消防员进入危险地区去营救事故受害者，通常委派给登高车连。

- **Rescue life on the fire ground** /ˈreskjuː - laɪf - ɒn - ðə - ˈfaɪə(r) - graʊnd/: the operation of firefighters on the fire ground to rescue threatened persons by various means

火场救生：消防员在火场中通过各种手段营救受困人员的行动。

- **Residential sprinkler system/ˌrezɪˈdenʃl - ˈsprɪŋklə - ˈsɪstəm/:** a sprinkler system arranged for fire suppression in a dwelling
  住宅自动喷水灭火系统：在住宅内安装用于灭火的自动喷水灭火系统。

- **Residual pressure /rɪˈzɪdjuəl - ˈpreʃə(r)/:** the amount of pressure in a hydrant system when a hydrant is fully open, such as during a fire; it should be engineered to provide domestic supply of water to homes and businesses during a large fire in the district
  剩余压力：火灾期间，消防栓打开时所含的压力量；发生大火时能为家庭和企业提供生活用水的压力。

- **Resistance /rɪˈzɪstəns/:** a force which slows down a moving object or vehicle
  阻力：阻止物体移动或车辆速度的力。

- **Response time /rɪˈspɒns - taɪm/:** the time from receiving a fire or other emergency signal to the time when the fire truck leaves the fire station

  响应时间：从收到火灾或其他紧急信号到消防车离开消防站的时间。

- **Response to fire alarm /rɪˈspɒns - tu - ˈfaɪə(r) - əˈlɑːm/:** the activity of receiving and processing the reported fire alarm information through various channels and methods

  火警响应：通过各种渠道和方法接收和处理上报的火警信息的活动。

- **Reverse lay /rɪˈvɜːs - leɪ/:** the process of stringing hose from a fire toward a source of water, e.g., a fire hydrant

  反向铺设：将软管从火中拉向水源（例如消防栓）的过程。

- **Rollover /ˈrəʊləʊvə(r)/:** the ignition of ceiling-level fire gases

  滚燃：达到上限标准的火灾气体的燃烧。

S

- **Safety curtain /ˈseɪfti - ˈkɜːtn/:** a movable curtain that prevents the smoke and heat generated by the fire from passing through

  防火幕： 阻止火灾产生的烟气和热量通过的可移动式幕布。

- **Salvage /ˈsælvɪdʒ/, salvage cover/ˈsælvɪdʒ - ˈkʌvə(r)/:** the heavy-duty tarpaulins folded or rolled for quick deployment to cover personal property subjected to possible water or other damage during fire fighting

抢救、抢救盖布：可快速展开或折叠的重型防水油布，在灭火时用于遮盖可能遭受水或其他损害的个人财产。

- **Scene safety /siːn - ˈseɪfti/:** the steps taken at or near an emergency scene to reduce hazards and prevent further injuries to workers, victims or bystanders

  现场安全：在紧急事故现场或附近为减少危险或防止对工人、受害者或旁观者造成进一步伤害而采取的措施。

- **SCBA:** the Self Contained Breathing Apparatus (SCBA), which means you have your oxygen tank and mask, keeping you from breathing in smoke or hazardous gases, as a part of your personal protective equipment (PPE)

  自给式呼吸器：可防止吸入烟气或有害气体的氧气瓶和面罩，是个人防护装备 (PPE) 的一部分。

- **Scattered receipt of fire alarm /ˈskætəd - rɪˈsiːt - əv - ˈfaɪə(r) - əˈlɑːm/:** the fire brigade directly accepts the report of fire alarm in the jurisdiction

分散接警：消防队直接接受辖区内的火灾报警报告。

- **Sealed automatic sprinkler system /si:ld - ˌɔːtəˈmætɪk - ˈsprɪŋklə - ˈsɪstəm/:** an automatic sprinkler system using sealed sprinkler head, including automatic wet sprinkler system, automatic dry sprinkler system, and pre-action automatic sprinkler system, etc.

  闭式自动喷水灭火系统：采用闭式洒水喷头的自动喷水灭火系统，其中包括湿式自动喷水灭火系统、干式自动喷水灭火系统、预作用自动喷水灭火系统等。

- **Sector /ˈsektə/**

  区段

- **Self-extinguishing ability /self-iksˈtiŋgwiʃiŋ - əˈbɪləti/:** a material terminates combustion after the ignition source is removed under specified test conditions

  自熄性：在规定的测试条件下，材料在移除火源后停止燃烧。

- **Shaft /ʃɑːft/:** a long, narrow, usually vertical passage in a building or underground, used for a lift/elevator or as a way of allowing air in or out

  （电梯的）升降机井；通风井；竖井：建筑物或地下狭长的、通常为垂直状的通道，作为电梯或使空气进出。

- **Sides /saɪdz/ A, B, C, and D:** the terms used by firefighters labeling the multiple sides of a building starting with side A or Alpha being the front of the structure and working its way around the outside of the structure in a clockwise direction. This labels the front side A or Alpha, the left side B or Bravo, the rear side C or Charlie, and the right side D or Delta.

  A、B、C 和 D 面：消防员使用的术语，用于标记建筑物的侧面。建筑物的前部为 A 面或 Alpha 面，再沿顺时针方向围绕建筑物外部进行标记，左侧为 B 面或 Bravo 面，背面为 C 面或 Charlie 面，右侧为 D 面或 Delta 面。

- **Situation awareness /ˌsɪtjuˈeɪʃən - əˈwɛənis/:** the perception of environmental elements with respect to time and/or space. It is a field of study concerned with perception of the environment critical to decision-makers in complex, dynamic areas from aviation, air traffic control, ship navigation, power plant operations, military command and control, and emergency services such as fire fighting and policing

  态势感知：对时间和（或）空间方面的环境要素的感知。它是一个研究领域，涉及对复杂、动态领域的重要感知，这些领域包括航空，空中交通管制、船舶导航、发电厂运营、军事指挥和控制以及消防和警务等应急服务。

- **Size-up /saɪz-ʌp/:** an initial evaluation of an incident, in particular a determination of immediate hazards to responders, other lives and property, and what additional resources may be needed. Example: "Two-story brick taxpayer with heavy smoke showing from rear wooden porches and children reported trapped."

评估：对事故进行初步评估，特别是确定对爱国人员、其他生命和财产的直接危害，以及可能需要的额外资源。示例："两层砖房，后木门廊冒出浓烟，据报有儿童被困。"

- **Sliding pole /ˈslaɪdɪŋ - pəʊl/:** a cylindrical pole in the fire station for firefighters to slide directly from a height to a designated position

  消防滑杆：消防站中的圆柱形杆，供消防员直接从高处滑到指定位置。

- **Smoke /sməʊk/:** the airflow formed by solid and liquid particles and gases produced during the pyrolysis or combustion of a substance, together with some mixed air

  烟（气）：由物质热解或燃烧产生的固体和液体颗粒、气体和混合空气一起形成的气流。

- **Smoke bay /sməʊk - beɪ/:** a local space separated by smoke blocking facilities inside the building, which can prevent fire smoke from spreading to the rest of the same building within a certain period of time

防烟分区：由建筑物内的防烟设施隔开的局部空间，防止火灾产生的烟气在一定时间内蔓延到本建筑物的其他地方。

- **Smoke control system /sməʊk - kənˈtrəʊl - ˈsɪstəm/:** a system that uses mechanical pressurization or natural ventilation to prevent smoke from entering stairwells, front rooms, refuge floors (rooms) and other spaces

  烟气控制系统：采用机械加压或自然通风方式来防止烟气进入楼梯间、前室、避难层（间）和其他空间的系统。

- **Smoke extraction system /sməʊk - ɪkˈstrækʃ(ə)n - ˈsɪstəm/:** a system that uses mechanical smoke extraction or natural smoke control to exhaust smoke to the outside of a building

  排烟系统：采用机械排烟或自然排烟方式将烟气排至建筑物外部的系统。

- **Smoke extinguishing system /smǝʊk - iksˈtiŋgwiʃiŋ - ˈsɪstǝm/:** a fire extinguishing system in which the smoke fire extinguishing agent undergoes combustion reaction in the smoke fire extinguisher to produce smoke extinguishing gas, which is sprayed above the fire liquid surface in the storage tank to form a uniform and thick fire extinguishing gas layer

  烟雾灭火系统：烟雾灭火剂在烟雾灭火器内发生燃烧反应，产生灭烟气体，喷洒在储罐的火液表面上方，形成均匀厚实的灭火气层的灭火系统。

- **Smoke layer /smǝʊk - ˈleɪǝ(r)/:** caused by a fire, a relatively uniform amount of smoke, which is caused by a fire, formed and accumulated under the highest interface of the enclosed space

  烟（气）层：由火灾引起，在封闭空间的最高界面下形成并积累相对均匀量的烟气。

- **Smoke management system /smǝʊk - ˈmænɪdʒmǝnt - ˈsɪstǝm/:** the general term for the smoke control system and smoke extraction system that is installed in a building to prevent the spread of fire smoke

烟气管理系统：安装在建筑物中的烟气控制系统和排烟系统的总称，用于防止火灾烟气蔓延。

- **Smoke proof staircase /sməʊk - pruːf - ˈsteəkeɪs/:** the installed anti-smoke front rooms, open balconies or recessed corridors and other facilities (collectively called front rooms) at the entrance of a stairwell to prevent fire smoke and heat from entering the stairwell

防烟楼梯：在楼梯间入口处安装的防烟前室、开放式阳台或嵌入式走廊等设施（统称前室），以防止火灾的烟气和热量进入楼梯间。

- **Smoke stratification /sməʊk - ˌstrætɪfɪˈkeɪʃən/:** the layered state of flue gas caused by thermal effect in an enclosed space without air flow disturbance

烟气分层：在没有气流扰动的封闭空间中由热效应引起的烟气分层状态。

- **Smouldering** /ˈsmǝʊldǝrɪŋ/: slow combustion of a substance without visible light, usually resulting in smoke and an increase in temperature

  阴燃：没有可见光的物质缓慢燃烧，通常会导致烟气和温度升高。

- **Solid** /ˈsɒlɪd/: a substance or an object that is hard or firm, not in the form of a liquid or gas

  固体：固体物质或物体，而非液体或气体。

- **Solid stream** /ˈsɒlɪd - striːm/: fire stream from round orifice of nozzle, different from straight stream

  密集水流：喷嘴圆形孔口射出的消防射流，与"直流"不同。

- **Soot** /sʊt/: black power or particles produced and deposited by the incomplete combustion of organic matter, mainly carbon particles

  烟灰：由有机物不完全燃烧产生和沉积的颗粒，主要是碳颗粒。

- **Specific toxicity** /spəˈsɪfɪk - tɒkˈsɪsəti/: toxicity caused by exposure to a poison that causes mutagenesis, teratogenesis, carcinogenesis, sensitization to organisms
特定毒性：因暴露于导致生物体突变、致畸、致癌、致敏的毒物引起的毒性。

- **Spontaneous ignition** /spɒnˈteɪniəs - ɪgˈnɪʃ(ə)n/: the combustion of combustible materials caused by heating or self-heating and accumulating heat without external fire source
自燃：可燃物在没有外部火源的情况下，因加热或自加热并积聚热量而引起的燃烧。

- **Spontaneous ignition temperature** /spɒnˈteɪniəs - ɪgˈnɪʃ(ə)n - ˈtemp(ə)rətʃə/: the lowest temperature at which combustibles can spontaneously ignite under specified conditions
自燃温度：可燃物在特定条件下可以自燃的最低温度。

- **Sprinkler** /ˈsprɪŋklə/: a device used to spray water

喷水灭火装置：用于喷水的装置。

- **Sprinkler-foam extinguishing system /ˈsprɪŋklə -fəʊm - iksˈtɪŋgwiʃɪŋ - ˈsɪstəm/:** an automatic sprinkler system that can spray water and foam is formed after the equipment for supplying the foam mixture is configured
自动喷水 – 泡沫联用灭火系统：配置供应泡沫混合液的设备后，形成可以喷水和泡沫的自动喷水灭火系统。

- **Stack effect /stæk - ɪˈfekt/:** the movement of air into and out of building, chimneys, flue gas stacks, or other containers, and is driven by buoyancy. Buoyancy occurs due to a difference in indoor-to-outdoor air density resulting from temperature and moisture differences
烟囱效应：空气进出建筑物、烟囱、烟前气烟囱或其他容器的运动，由浮力推动。浮力的产生是由于温度和湿度的差异导致的室内外空气密度的差异。

消防救援英语关键词

- **Staging /ˈsteɪdʒɪŋ/:** sector of incident command where responding resources arrive for assignment to another sector, often an essential element in *personnel accountability* program

  集结待命区：事故指挥的一个区段，救援资源到达集结区，然后分配给其他区段，通常是火场责任系统的基本要素。

- **Standard Duration Breathing Apparatus (SDBA) sets/ˈstændəd - djuˈreɪʃn - ˈbriːðɪŋ - ˌæpəˈreɪtəs - sets/:** a type of frogman's rebreather breathing set. Many of the world's navies and army marine corps have used it since 1971

  标准持续时间呼吸器套装：一种蛙人换气呼吸器，1971年世界上许多海军和陆军战队都用过。

- **Standard operating procedure /guideline (SOP or SOG) /ˈstændəd - ˈɔpəˌreɪtɪŋ - prəˈsiːdʒə(r) - ˈgaɪdlaɪn/:** rules for the operation of a fire department, such as how to respond to various types of emergencies,

158

training requirements, use of protective equipment, radio procedures; often include local interpretations of regulations and standards. In general, "procedures" are specific, whereas "guidelines" are less detailed.

标准操作程序、指南（SOP 或 SOG）：消防部门的运行规则，例如如何应对各种类型的紧急情况、培训要求、防护设备的使用、无线电程序；通常包括当地对法规和标准的解释。一般而言，"程序"更加具体详细，而"指南"对于细节描述较少。

- **Standard time-temperature curve** /ˈstændəd - taɪm - ˈtemp(ə)rətʃə - kɜːv/: a function curve in which the temperature in the fire test furnace changes with time in a standard fire test process

标准时间 - 温度曲线：在标准防火测试过程中，防火试验炉内的温度随时间变化的函数曲线。

- **Starting for fire fighting** /ˈstɑːtiŋ - fɔ - ˈfaɪə(r) - ˈfaɪtɪŋ/: firefighters wear fire fighting clothing and take fire trucks, boats or planes to the fire ground, after receiving the dispatch order

灭火出动：接到调度命令后，消防员穿上消防服，乘坐消防车、船只或飞机前往火场。

- **Static pressure /ˈstætɪk - ˈpreʃə(r)/:** the pressure in a water system when the water is not flowing

  静压：水不流动时水系中的压力。

- **Straight stream /streɪt - striːm/:** round, hollow stream formed as water passes a round baffle through a round orifice (e.g., on an adjustable nozzle), different from solid stream

  直流：当水通过圆形孔口（例如，在可调节喷嘴上）流过圆形挡板时形成的圆形空心流，与"密集水流"不同。

- **Stretch /stretʃ/:** command to lay out (and connect) fire hose and nozzle

  展设水带：展开（和连接）消防水带和喷嘴的命令。

- **Structure fire /ˈstrʌktʃə(r) - ˈfaɪə(r)/ (or structural fire):** a fire in a residential or commercial building. Urban fire departments are primarily geared toward structural fire fighting. The term is often used to distinguish them from wildland fire or other outside fire, and may also refer to the type of training and equipment such as "structure PPE" (personal protective equipment)

  建筑火灾: 住宅或商业建筑中的火灾。城市消防部门主要面向建筑灭火。该术语通常用于把建筑火灾与野外火灾或其他外部火灾区分开来，也可以指培训和设备的类型，例如"建筑 PPE"（个人防护设备）。

- **Sworn personnel /swɔːn - ˌpɜːsəˈnel/:** firefighters take a sworn oath to protect and serve the community in which they work

  宣誓人员: 宣誓保护和服务社会的消防员。

- **Switch /swɪtʃ/:** a small device that you press or move up and down in order to turn a light or a piece of electrical equipment on and off

（电路的）开关，闸：通过上下按压或移动来打开或关闭灯，电器的小型装置。

T

- **Tachometer /tæˈkɒmɪtə/:** an instrument measuring the rotation speed of a shaft or disk, as in a motor or other machine; RPM (revolutions per minute) gauge

  转速计，流速计：汽车或其他机器中用于测量轴或圆盘转速的仪器；每分钟转数计量器。

- **Tailboard /ˈteilbɔ:d/:** a portion at rear of fire engine where firefighters could stand and ride (now considered overly dangerous and against department policy in Carmel), or step up to access hoses in the hose bed

尾板：消防车后边的部分，消防员可以站立和骑坐的位置（现在认为过于危险并违反卡梅尔的部门政策）；或可经此获取水带箱中的软管。

- **Tanker /ˈtæŋkə(r)/:** an aircraft equipped to carry water or fire retardant for use in wildland fire suppression
  灭火飞机：载水或常有阻燃剂的飞机，用于野外灭火。

- **Tetrahedron /ˌtetrəˈhiːdrən/:** a solid shape with four flat sides that are triangles
  四面体：有四个平面三角形的立体形状。

- **The cleveland coil /ðə - ˈklivlənd - kɔɪl/**
  （水带）克利夫兰卷（主要用于高层建筑灭火）

- **Thermal /ˈθɜːml/:** connected with heat
  热的；热量的：与热量相关。

- **Thermal Imaging Camera (TIC) /ˈθɜːml - ˈɪmɪdʒɪŋ - ˈkæm(ə)rə/:** a device that forms an image using infrared radiation, similar to common camera that forms and image using visible light

热像仪：利用红外辐射成像的设备，类似于普通摄像机使用可见光成像。

- **Thermometer /θəˈmɒmɪtə(r)/:** an instrument for measuring temperature

温度计：测量温度的仪器。

- **Transmit /trænzˈmɪt/:** to allow heat, light, sound, etc., to pass through

传（热、声等）；透（光等）

- **Trench effect /trentʃ - ɪˈfekt/:** a combination of circumstances that can rush a fire up an inclined surface

沟槽效应：由多种情况结合而成，可以使火势迅速蔓延到倾斜的表面。

- **Truck company /ˈtrʌk - ˈkʌmpəni/:** a group of firefighters assigned to an apparatus that carries ladders, forcible entry tools, possibly extrication tools and salvage covers, and otherwise equipped to perform rescue, ventilation, overhaul and other specific functions at fires; also called "ladder company"

登高车中队 / 连队：配有梯子、破拆工具、解救工具和抢救盖布的消防员队，并配备在火灾中执行救援、通风、检修和其他特定功能的其他设备；又称"云梯车连"。

- **Turnout gear/ˈtɜːnaʊt - ɡɪə(r)/:** the protective clothing worn by firefighters

消防战斗服：消防员穿的防护服。

- **Two-in, two-out (or "two in/two out") /tuː - ɪn, tuː - aʊt/:** refers to the OSHA standard safety tactic of having one team of two firefighters enter a hazardous zone (IDLH), while at least two others stand by outside in case the first two need rescue — thus requiring a minimum of four firefighters on scene prior to starting interior attack. Also it refers to the "buddy system" in which firefighters never enter or leave a burning structure alone.

Note: OSHA stands for Occupation Safety and Health Administration which is under the Department of Labour of the USA

两进两出：根据 OSHA 标准安全策略，派两名消防员组成的团队进入危险区域 (IDLH)，同时至少两名其他消防员在外面待命，以防前两名消防员需要救援，因此，在开始内部攻击之前，至少需要四名消防员在现场；也指消防员绝不单独进入或离开燃烧建筑的"伙伴系统"。

注：OSHA 是美国职业安全和健康署，隶属美国劳工部。

- **Type /taɪp/ I, II, III, IV, V Building /ˈbɪldɪŋ/:** the USA classification system for fire resistance of building construction types, including definitions for "resistive" Type I, "non-combustible" Type II, "ordinary" Type III, "heavy timber" Type IV, and "frame construction" Type V (i.e., made entirely of wood).

  I、II、III、IV、V 型建筑：美国建筑结构类型的耐火性分类系统，包括"阻燃"I 型、"不可燃"II 型、"普通"III 型、"重木材"IV 型和"框架结构"V 型（即完全由木头制成）。

- **Thermal radiation /ˈθɜːml - reɪdɪˈeɪʃ(ə)n /:** heat energy transferred in the form of electromagnetic waves
  热辐射： 以电磁波形式传递的热能。

- **Total flooding extinguishing system /ˈtəʊt(ə)l - ˈflʌdɪŋ - iksˈtiŋgwiʃiŋ - ˈsɪstəm/:** a fixed fire extinguishing system that fills the protected enclosed space with a certain concentration (strength) of extinguishing agent (gas, high-expansion foam, etc.) to achieve the purpose of extinguishing fire
  全淹没灭火系统： 一种固定式灭火系统，在受保护的封闭空间内填充一定浓度（强度）的灭火剂（气体、高倍数泡沫等），以达到灭火目的。

- **Total flooding of high expansion foam extinguishing system /ˈtəʊt(ə)l - ˈflʌdɪŋ - əv - haɪ - ɪkˈspænʃn - fəʊm-**

iks'tiŋgwiʃiŋ - 'sistəm/: a foam extinguishing system in which high expansion foam is sprayed by a stationary high expansion foam generator into an enclosed protective zone and reaches the submerged depth within a specified time

全淹没式高倍数泡沫灭火系统：固定式高倍数泡沫发生装置将高倍数泡沫喷射到封闭的防护区，并在规定时间内达到淹没深度的泡沫灭火系统。

- **Toxic hazard /'tɒksɪk - 'hæzəd/:** the harmful effects on organisms caused by the production of toxicant in a fire

  有毒危害：火灾中产生的有毒物质对生物体造成的有害影响。

- **Toxic model /'tɒksɪk - 'mɒdl/:** a device for evaluating the toxicity of materials in a fire under specified test conditions

  毒性模型：在规定的测试条件下，评估材料在火灾中的毒性的装置。

- **Toxic potency /ˈtɒksɪk - ˈpəʊtnsɪ/:** the intensity of the harmful biological changes produced by the toxicant

  毒性效力：有毒物质使有害生物产生变化的强度。

- **Toxic risk /ˈtɒksɪk - rɪsk/:** possibility of toxic hazard in a fire

  毒害风险：火灾中产生有毒危害的可能性。

- **Toxicant /ˈtɒksɪk(ə)nt/:** the substances capable of causing toxicity to organisms

  毒物：能够使生物体中毒的物质。

- **Toxicant concentration /ˈtɒksɪk(ə)nt - ˌkɒnsnˈtreɪʃn/:** the content of toxicant per unit volume of air

  Note: It is usually expressed in terms of mass concentration (mg/L or g/m^3) or volume fraction (10%—6%).

  毒物浓度：单位体积空气中的毒物含量。

  注: 通常以质量浓度 (mg/L 或 g/m^3) 或体积分数 (10%—6%) 表示。

170

- **Toxicant dose /ˈtɒksɪk(ə)nt - dəʊs/:** the amount of poison inhaled by an organism

  Note: In toxicology, the toxicant dose can be determined by multiplying the exposure dose by the average volume of air inhaled by an organism per unit time, expressed in mg/min.

  毒物剂量： 生物体所吸入的毒物量。

  注： 在毒理学中，毒物剂量可用暴露剂量乘以单位时间生物体吸入空气的平均体积来确定，单位为 mg/min。

- **Toxicity /tɔkˈsisəti/:** the property of a substance that has harmful effects on living organisms

  毒性： 物质对生物体产生有害影响的特性。

- **Turbulent flame /ˈtɜːbjələnt - fleɪm/:** the flame that exhibits irregular flow when burning

  湍流火焰： 燃烧时表现出不规则流动的火焰。

- **Turnout alert bell /ˈtɜːnaʊt - əˈlɜːt - bel/:** the sound equipment that can convey the fire and rescue order after receiving an alarm

出动警铃: 收到警报后传达消防救援命令的音响设备。

- **Turnout alert lamp** /ˈtɜːnaʊt - əˈlɜːt - læmp/: a lamp that can display the fire fighting and rescue command and send out an optical warning signal after receiving an alarm

  出动警灯: 接警后，能显示消防救援命令并发出光学警告信号的灯。

U

- **Under control /ˈʌndə - kənˈtrəʊl/:** the fire or spill etc., is no longer spreading. The situation is contained. This term should not be confused with a report that the fire is out.

  受到控制： 火灾等不再蔓延，局势得到控制。该术语与"火灾已扑灭"含义不同，不要混淆。

- **Uniform /ˈjuːnɪfɔːm/:** not varying; the same in all parts and at all times

  一致的；统一的： 不变；所有部分在任何时候都相同。

- **United States Fire Administration (USFA) /juˈnaɪtɪd-steits - ˈfaɪə(r) - ədˌmɪnɪˈstreɪʃn/:** a division of the Federal Emergency Management Agency (FEMA), which is managed by the Department of Homeland Security (DHS) of the USA

  美国消防局 (USFA)：联邦紧急事务管理局 (FEMA) 的一个部门，该部门由国土安全部 (DHS) 管理。

- **Universal precautions /ˌjuːnɪˈvɜːsl - prɪˈkɔːʃnz/:** the use of safety barriers (gloves, masks, goggles) to limit an emergency responder's contact with contaminants, especially fluids of injured patients

  通用预防措施：使用安全屏障（手套、面罩、护目镜）来限制应急人员接触污染物，尤其是受伤患者的血液。

- **Utility truck /juːˈtɪləti - trʌk/:** usually manned by an engine company and responds to utility calls like water main breaks. Some small departments of the USA use them to respond to medical calls on gas saving.

多功能消防车：通常由消防队驾驶，并处理水管破裂等公用设施故障。在美国，一些小部门用它们来回应医疗呼叫以节省汽油费。

V

- **Vapour** /ˈveɪpə/: a mass of very small drops of liquid in the air, such as steam

  蒸气： 空气中大量非常小的液滴，例如蒸汽。

- **Velocity** /vəˈlɒsəti/: the speed of something in a particular direction

  （沿某一方向的）速度

- **Vehicle fire** /ˈviːəkl - ˈfaɪə(r)/: the type of fire involving motor vehicles themselves, their fuel or cargo; has peculiar issues of rescue, explosion sources, toxic smoke and runoff, and scene safety

车辆火灾：涉及机动车辆本身、其燃料或货物的火灾类型；在救援、爆炸源、有毒烟雾和径流、现场安全等方面存在特有的问题。

- **Ventilation /ˌventɪˈleɪʃ(ə)n/:** an important procedure in fire fighting in which the hot smoke and gases are removed from inside a structure, by either natural or forced convection, and through either existing openings or new ones provided by firefighters at appropriate locations (e.g., on the roof). Proper ventilation can save lives and improper ventilation can cause backdraft or other hazards.

通风：消防中的重要程序，通过自然或强制对流，以及通过现有通风口或消防员在适当位置（如屋顶）提供的新通风口，从建筑内部排出热烟和气体。适当的通风可以挽救生命，通风不当会导致回燃或其他危险。

- **Venturi effect /venˈtjuəri - ɪˈfekt/:** creating a partial vacuum using a constricted fluid flow, used in fire equipment for mixing chemicals into water streams, or for measuring flow velocity

  文丘里效应：利用收缩的流体流动产生部分真空，用于消防设备，将化学品混合到水流中，或用于测量流速。

- **Vertical ventilation /ˈvɜːtɪkl - ˌventɪˈleɪʃ(ə)n/:** ventilation technique making use of the principle of convection in which heated gases naturally rise

  垂直通风：利用加热气体自然上升的对流原理形成的通风技术。

- **Vessel /ˈvesl/:** a container used for holding liquids

  （盛液体的）容器，器皿

- **Voids /vɔɪdz/ (building):** the enclosed portions of a building where fire can spread undetected

  空隙（建筑物）：建筑物的封闭部分，火灾可以在不被发现的情况下蔓延。

- **Volunteer fire brigade/ˌvɒlənˈtɪə(r) - ˈfaɪə(r) - brɪˈɡeɪd/:** the fire brigades established by government agencies, organizations, enterprises, institutions, and township people's governments, villagers' committees and residents' committees to undertake self-prevention and self-rescue work

  志愿消防队：由政府机构、组织、企事业单位和乡镇人民政府、村民委员会、居民委员会设立的消防队，开展自卫自救工作。

- **Volt /vəʊlt/:** a unit used to measure the force of an electric current

  伏特（电压单位）：测量电流强度的单位。

W

- **Water hammer /ˈwɔːtə - ˈhæmə(r)/:** the large, damaging shock wave in a water supply system caused by shutting a valve quickly, or by permitting a vehicle to drive across an unprotected fire hose

  水锤现象：由于快速关闭阀门或车辆驶过未受保护的消防水带导致供水系统产生大型破坏性冲击波。

- **Water spray extinguishing system /ˈwɔːtə - spreɪ - iksˈtiŋgwiʃiŋ - ˈsɪstəm/:** an open automatic sprinkler system, composed of water source, water supply equipment, pipeline, rain alarm valve, filter and water mist nozzle, etc., sprays water mist to extinguish fire or protects and cools the object to be protected

水喷雾灭火系统：一种开放式自动喷水灭火系统，由水源、供水设备、管道、雨水报警阀、过滤器和水雾喷头等组成，通过喷洒水雾灭火或保护和冷却物体。

- **Water supply to fire ground /ˈwɔːtə - səˈplaɪ - tu - ˈfaɪə(r) - graʊnd/:** the operation of delivering water to a fire ground by using fire trucks, fire boats, fire pumps and other fire supply appliances

火场供水：使用消防车、消防船、消防泵和其他消防设备向火场供水的操作。

- **Watt /wɒt/:** a unit for measuring electrical power

瓦，瓦特：电功率单位。

- **Well involved /wel - ɪnˈvɒlvd/:** a term of size-up meaning fire, heat and smoke in a structure are so widespread that internal access must wait until fire streams can be applied

充分参与：规模扩大的术语，指建筑中的火、热量和烟雾大范围蔓延，以至于必须等消防火流解决后才能进入建筑物内部。

- **Wet automatic sprinkler system /wet - ˌɔːtəˈmætɪk - ˈsprɪŋklə - ˈsɪstəm/:** in the quasi-working state, the water distribution pipeline is filled with pressurized water to start the system with a sealed automatic sprinkler system
  湿式自动喷水灭火系统：准工作状态下，配水管道内充满用于启动系统的有压水的闭式自动喷水灭火系统。

- **Wet down ceremony /wet - daʊn - ˈserəməni/:** a traditional American fire ceremony from the USA for the placing of new apparatus in service. There are several versions of this but it usually includes: pushing the old apparatus out, wetting down the new vehicle and pushing it back into the station. It may also include the moving of the bell to the new apparatus, photos, etc.

洗礼仪式: 美国新设备投入使用的美国消防传统仪式，该仪式有多种版本，通常包括：推出旧设备，浇湿新车并推回消防站，将消防警铃移动到新设备旁边拍照等。

- **Windward /ˈwɪndwəd/:** a term to describe the side of something, especially a ship, which is facing the wind
  迎风面: 形容物体的某一面，尤其是指船迎风的一面。

- **Wired alarm /ˈwaiəd - əˈlɑ:m/:** the report of fire alarm through wired fire communication network
  有线报警: 通过有线消防通信网络进行火灾报警。

- **Wired fire communication line /ˈwaiəd - ˈfaɪə(r) - kəˌmjuːnɪˈkeɪʃn - laɪn/:** a wired communication line used to transmit fire alarm and fire information
  有线消防通信线路: 用于传输火灾报警信息的有线通信线路。

- **Wired fire communication network /ˈwaiəd - ˈfaɪə(r) -

kə ˌmjuːnɪˈkeɪʃn - ˈnetwɜːk/: a communication network composed of wired fire communication equipment and wired fire communication lines

有线消防通信网络：由有线消防通信设备和有线消防通信线路组成的通信网络。

- **Wireless alarm /ˈwaɪələs - əˈlɑːm/:** the report of fire alarm through wireless fire communication network

无线报警：通过无线消防通信网络进行火灾报警。

- **Wireless fire communication network /ˈwaɪələs - ˈfaɪə(r) - kə ˌmjuːnɪˈkeɪʃn - ˈnetwɜːk/:** in a certain communication area, a fire fighting wireless communication network composed of wireless communication equipment and necessary communication channels

无线消防通信网络：在一定的通信区域内，由无线通信设备和必要的通信信道组成的无消防线通信网络。

- **Withdrawal /wɪðˈdrɔːəl/:** an action to pull or take somebody

/ something back or away

撤退

- **Wreckage /ˈrekɪdʒ/:** the remains of something that has been badly damaged or destroyed

（遭损毁之物的）残骸

Z

- **Zone /zəʊn/:** the section of structure indicates on fire alarm control panel where sensor is activated

  区域：火灾报警控制面板上的传感器被激活的部分。

参考文献

References

[1] *Carmel Clay TWP. Fire Department Terminology*

[2] The Institntion of Fire Engineer. Elementary Fire Engineering Handbook (2016)

[3] GB/T5907.1—2014 消防词汇 第 1 部分：通用术语
GB/T5907.1—2014 Fire Protection Vocabulary Part 1: General Terms

[4] GB/T 4968—2008 火灾分类
GB/T 4968—2008 Fire classification

[5] GB/T5907.2—2015 消防词汇 第 2 部分：火灾预防
GB/T5907.2—2015 Fire Protection Vocabulary Part 2: Fire Prevention

[6] GB/T 5332—2007 可燃液体和气体引燃温度试验方法
GB/T 5332—2007 Test method for ignition temperature of combustible liquid and gas

[7] GB/T 5907.3—2015 消防词汇 第 3 部分：灭火救援
GB/T 5907.3—2015 Fire Protection Vocabulary Part 3: Fire Fighting and Rescue

[8] GB/T 5907.5—2015 消防词汇 第 5 部分：消防产品

GB/T 5907.5—2015 Fire Protection Vocabulary Part 5: Fire Products

[9] GB 13495—1992 消防安全标志

GB 13495—1992 Fire Protection Safety Signs

[10]《中华人民共和国消防法》(2008 年发布）

Fire Protection Law of the People's Republic of China (promulgated in 2008)

[11] ISO 8421-1: 1987 Fire protection—Vocabulary—Part 1: General terms and phenomena of fire

[12] ISO 8421-2: 1987 Fire protection—Vocabulary—Part 2: Structural fire protection

[13] ISO 8421-5: 1988 Fire protection—Vocabulary—Part 5: Smoke control

[14] ISO 8421-6：1987 Fire protection—Vocabulary—Part 6: Evacuation and means of escape

[15] ISO 13943：2008 Fire safety—Vocabulary

附录

消防救援有关单位、职务英文译法 [①]

单　位		职　务		备注
中文	英文	中文	英文	
国家消防救援局	National Fire and Rescue Administration	局长	Director	
		政委	Political Commissar	
消防救援总队	Fire and Rescue Corps	总队长	Corps Commander	
支队	Detachment	支队长	Head of the Detachment	
大队 /	Brigade	大队长	Brigade Chief	
营	Battalion	营长	Battalion Chief	美国
中队	Detachment	中队长	Detachment Chief	
消防站	Fire Station / Firehouse	站长	Fire Station Chief / Station Manager	
消防连 / 小组	Fire Company	组长	Fire Company Leader	美国
党委	Party Committee	书记	Secretary	
		政委	Political Commissar	

[①] 参考：百度文库，网址：https://wenku.baidu.com/view/b4b0cbedd6bbfd0a 79563c1ec5da50e2524dd1f5.html。机构设置信息截至 2023 年 5 月 1 日。

续表

单　位		职　务		备　注
中文	英文	中文	英文	
办公室	General Office	主任	Director	
组织教育处	Organization and Moral Education Section	处长	Director	
人事处	Division of Human Resource	处长	Director	
队务处	Firefighters Management Division	处长	Director	
政策研究处	Policy Research Division	处长	Director	
防火监督处	Fire Prevention and Supervision Division	处长	Director	
战训处	Operation & Training Division	处长	Director	
科技处	Science & Technology Division	处长	Director	
宣传处	Publicity Division	处长	Director	
后勤装备处	Logistics & Equipment Division	处长	Director	
审计处	Internal Audit Division	处长	Director	
财务处	Finance Division	处长	Director	
后勤部	Logistics Division	部长	Director	
秘书处	Secretariat	处长	Chief	
通信处	Communications Division	处长	Chief	
调度指挥中心	Dispatch and Command Center	主任	Chief	

单 位		职 务		备 注
中 文	英 文	中 文	英 文	
（监察）纪检保卫处	(Supervision) Discipline Inspection and Security Division	处长	Chief	
车管处	Vehicle Management Division	处长	Chief	
法规处	Legal Affairs Division	处长	Chief	
火调处	Fire Investigation Division	处长	Chief	
危管处	Hazardous Materials Supervision Division	处长	Chief	缩写：HAZMAT Supervision Division
教导大队	Fire Protection Training Unit	大队长	Chief	
		教导员	Political Instructor	
天津消防研究所	Tianjin Fire Research Institute	所长	Director	其他各消防研究所替换地区名称即可。
		党委书记	Secretary of CPC Committee of the Institute	
国家固定灭火系统和耐火构件质量监督检验中心	National Center for Quality Supervision and Test of Fixed Fire-Extinguishing Systems and Fire-Resisting Building Components	主任	Director	

单 位		职 务		备 注
中文	英文	中文	英文	
国家消防装备质量监督检验中心	National Center for Quality Supervision and Test of Fire Fighting Equipment	主任	Director	原译名：National Fire Equipment Quality Supervision Testing Center
国家消防电子产品监督检验中心	National Center for Quality Supervision and Test of Fire Electric Products	主任	Director	原译名：National Supervision and Test Centre for Fire Electronic Product Quality
国家防火建筑材料监督检验中心	National Center for Quality Supervision and Test of Fire-proof Building Materials	主任	Director	原译名：National Center for Quality Supervision and Testing of Fire Building Materials

续表

单　位		职　务		备注
中文	英文	中文	英文	
中国消防协会	China Fire Protection Association	理事长	President	中国消防协会缩写：CFPA
		秘书长	Secretary General	
		政委	Political Commissar	
中国消防救援学院	China Fire and Rescue Institute	院长	Dean	
消防工程系	Department of Fire Engineering	主任	Director 或 Coordinator	
消防指挥系	Department of Fire Commanding	主任	Director 或 Coordinator	
秘书	Secretary	教授	Professor	
参谋	Staff Officer	副教授	Associate Professor	
干事	Political Officer	讲师	Lecturer	或 Instructor（美语常用）
助理员	Logistical Assistant	助教	Teaching Assistant	

消防救援衔英汉对照

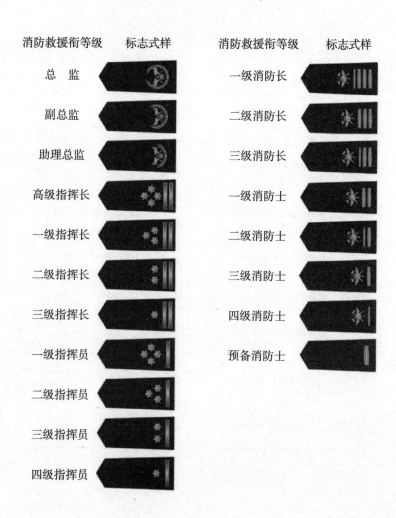

消防救援衔等级	标志式样	消防救援衔等级	标志式样
总　监		一级消防长	
副总监		二级消防长	
助理总监		三级消防长	
高级指挥长		一级消防士	
一级指挥长		二级消防士	
二级指挥长		三级消防士	
三级指挥长		四级消防士	
一级指挥员		预备消防士	
二级指挥员			
三级指挥员			
四级指挥员			

Collar Badge of Fire and Rescue Team of the PRC					
No.	Ranks	Markings	No.	Ranks	Markings
1	Commissioner		11	Fire Officer Fourth Class	
2	Deputy Commissioner		12	Senior Firefighter First Class	
3	Assistant Commissioner		13	Senior Firefighter Second Class	
4	Senior Fire Chief		14	Senior Firefighter Third Class	
5	Fire Chief First Class		15	Firefighter First Class	
6	Fire Chief Second Class		16	Firefighter Second Class	
7	Fire Chief Third Class		17	Firefighter Third Class	
8	Fire Officer First Class		18	Firefighter Fourth Class	
9	Fire Officer Second Class		19	Probationary Firefighter	
10	Fire Officer Third Class				

报告词英汉对照

上课报告词

Words of Report before Class

当授课教员和全体学员进入教室，上课铃声响后，值班员起立，走到离教员 2 米的位置处，面向学员，下达"全体起立"口令，然后面向教员，报告"教员，上课人员集合完毕，应到实到 XX 人，请您指示！值班员史密斯学员。"

（如有学员缺席，报告词改为："教员，上课人员集合完毕，实到 XX 人，XX 人病假 / 事假，请您指示！值班员史密斯学员。"）

授课教员回答："上课！"值班员回答"是，教员！"然后转向学员，下达"坐下"口令。学员坐下，值班员回到座位，课程开始。

When the teacher and all cadets enter the classroom and the bell rings, the duty cadet stands up and leaves his/her seat for the position which is 2 meters away from the teacher in the front of the classroom, facing the class and giving order: *All rise*. Then he/she turns towards the teacher and reports: *Madam/Sir, all the cadets are presented for class. Request to take class. Cadet Smith.* (or *Madam/Sir, 50 cadets are presented for class with 2 absent for sick leave/other purposes. Request to take class. Cadet Smith.*)

The teacher replies: *Class begins*. The duty cadet should respond: *Yes, Madam/Sir*. And he/she turns towards the class, giving order: *Sit down*. Then the duty cadet goes back to his/her seat. And the class kicks off.

下课报告词

Words of Report after Class

当授课教员宣布"今天课程到此结束",值班员起立,走到离教员 2 米的位置处,面向学员,下达"全体起立"口令,然后转向教员,报告"教员,授课完毕,请您指示!值班员史密斯学员。"

授课教员回答："下课！"值班员回答"是，教员！"然后转向学员，下达"坐下"口令。待教员先行离开后，值班员组织学员返回或转到另一个教室准备其他课程。

When the teacher declares: *That's all for today/ Class is over*, the duty cadet leaves his/her seat for the position 2 meters away from the teacher in the front of the classroom, facing the class and giving order: *All rise*. Then he/she turns towards the teacher and reports: *Madam/Sir, request to get dismissed*. The teacher replies: *Dismissed*. The duty cadet should respond: *Yes, Madam/ Sir*. Then he/she turns towards the class, giving order: *Sit down*. And the duty cadet leads the class back or to transfer to another classroom after the teacher leaves.

有限空间救援器材名称

Kits for Rescue in Confined Space

扁带 (harness)

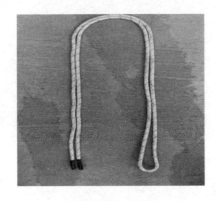

4 米绳 (4 m rope)

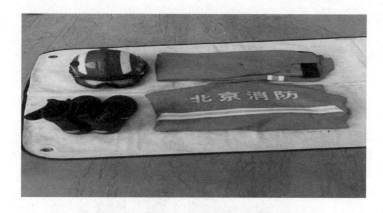

抢险服全套（帽子、上衣、裤子、抢险靴、腰带）
Rescue suit (fire hat, coat, trousers, rescure boots, safety belt)

救援三脚架 (Rescue tripod)

分力器 (rigging plate paw)
(a device used to divide the force)

下降器 (descender device)

绳索垫布 (rope pad)

单滑轮 (single pulley block)

6 米扁带 (6 m harness)

O 形钩 (O-shaped hook)

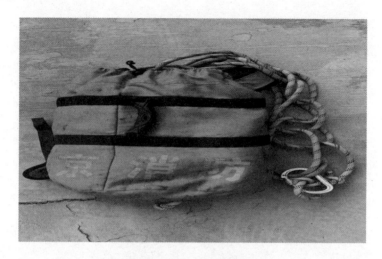

50 米绳包（10.5 毫米静力绳）
50 m rope bag (10.5 mm static rope)

全身吊带 (full body harness)

30 米绳 (30 m rope)

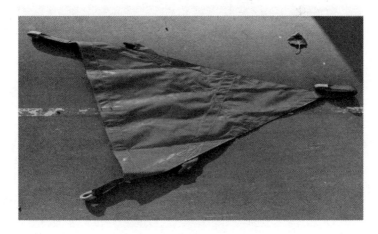

三角巾 (triangular scarf)

移动供气源 (mobile gas source)

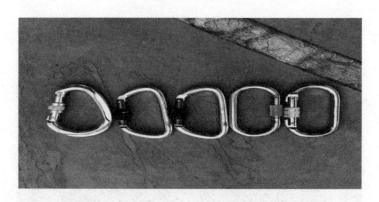

O 形钩 / D 形钩 (O-shaped hook / D-shaped hook)

19 件消防器材名称

序号	器材名称	备注	图片
1	20 式消防员灭火防护服（指挥款）	九江 ZFMH-JXA	
2	一级化学防护服	杜邦 Tychem TK554	
3	避火防护服	FBH-1	
4	正压式空气呼吸器	梅思安 AG2100	
5	消防员后场接收装置	河北永生 RHJ280T-C	
6	便携式移动照明灯	海洋王 FW6128	
7	手抬机动泵	东发泵 VC52AS	

续表

序号	器材名称	备注	图片
8	热成像仪	美国 FLIR K55	
9	救生照明线	河北永生 JS-1-B	
10	无齿锯	胡思货纳 K770 型	
11	移动消防炮	阿密龙 3443 型	
12	有毒气体检测仪	德尔格 X-am5000	
13	捆绑式堵漏袋	中正 KJ-21	
14	手持钢筋速断器	威盾 IS-MC19L	

续表

序号	器材名称	备注	图片
15	电动多功能工具	Strongarm E100	
16	可燃气体检测仪	Honey well Humidor Minimax X4	
17	音频生命探测仪	VIBRAPHONE ASB8	
18	雷达生命探测仪	RESQUE RADAR RQ-IV	
19	视频生命探测仪	盛博蓝 BF-V1000D	

1.Protective Clothing for Firefighters of the 20th Model

2.Class 1 Chemical Protective Clothing

3.Firemen Entry Clothing

4.Self-contained Positive Pressure Breathing Apparatus

5.Firemen Backfield Receiving Device

6.Portable Lighting Dimmer

7.Portable Fire Pump

8.Thermal Imager

9.Fire Rescue Lighting Line

10.Abrasive Disk Saw

11.Mobile Fire Cannon

12.Toxic Gas Detector

13.Strap-on Plugging Bag

14.Fire Rescue Rebar Cutter

15.Electric Multipurpose Tool

16.Combustible Gas Detector

17.Audio Life Detector

18.Radar Life Detector

19.Video Life Detector

消防救援常用术语（中英对照）

中文	英文
"全灾种、大应急"	"Comprehensive disasters and coordinated emergency response"
"一专多能"	"One Specialty, Multiple Functions"
"大安全、大应急"	"a framework of overall security and coordinated emergency response"
"一个班子、三支队伍"	"One party branch, three teams"
确保工作有人抓、有人管、有人干	The work is ensured to be organized, supervised, and carried out.
国家综合性消防救援队	The National Comprehensive Fire and Rescue Team
基层	the grassroots level

消防救援常用术语（中英对照）

英文	中文	例句
jaws of life	救生钳	Sometimes, the removal of a trapped victim during a vehicle extrication uses hydraulic spreader, jaws of life, or other technical equipments（实施交通事故人员解救过程中，有时使用液压扩张器、救生钳或其他技术设备。）
truck company	登高车连；中队	
guy line	绷绳	
zodiac	（配有尾挂式发动机的）（充气）橡皮艇	Search and rescue team comes with a zodiac（搜救队携橡皮艇前来救援。）